CONTEMPORARY ENVIRONMENTAL ISSUES

CONTEMPORARY ENVIRONMENTAL ISSUES

Judith Rosales

www.societypublishing.com

Contemporary Environmental Issues

Judith Rosales

Society Publishing

2010 Winston Park Drive,

2nd Floor

Oakville, ON L6H 5R7

Canada

www.societypublishing.com

Tel: 001-289-291-7705
001-905-616-2116
Fax: 001-289-291-7601
Email: orders@arclereducation.com

ISBN: 978-1-77407-307-0 (Softcover)

ABOUT THE AUTHOR

Prof. Judith is an Ecologist and Environmental Specialist with more than 30 years of academic and consulting experience. She holds a PhD in geography from the University of Birmingham UK and wrote the PhD and Master Academic Programs of Environmental Sciences for the Universidad de Guayana, Venezuela.

TABLE OF CONTENTS

LIST OF FIGURES

LIST OF ABBREVIATIONS

ARI	acute respiratory infection
CE	cost efficiency
CFCs	chlorofluorocarbons
CJD	Creutzfeldt-Jakob disease
CO_2	carbon dioxide
COPD	chronic obstructive pulmonary disease
CSP	concentrating solar thermal plants
DALYs	disability-adjusted life years
EE	energy efficiency
EM	environment management
GCMs	general circulation models
GHG	greenhouse gas
IAEA	International Atomic Energy Agency
IEQ	indoor environmental quality
kWh	kilowatt-hour
MDGs	millennium development goals
NAS	National Academy of Sciences
NOx	nitrogen oxides
NSP	nonpoint source pollution
NWCC	National Wildlife Conservation Council
ODS	ozone-depleting substances
PV	photovoltaic
SO_2	sulfur dioxide

TVA	Tennessee Valley Authority
UV	ultraviolet
WCED	World Commission on Environment and Development
WHO	World Health Organization

PREFACE

The environment is derived from the word "environ," which mean surroundings as forms a significant part of the world we the human species live in. It supports the very existence of the human beings and serves all the needs of human beings and other living beings that depend on it for their existence and survival.

The common public perception of the environment is that at the moment, it seems to be heavily affected in an adverse way by the activities of humans and other consequent damaging activities going on in the environment. Also that there are environmentalists who deal with several issues that surround the environment and its appropriate maintenance. People's perception rarely incorporates the need for their own change on behavior to support the actions of those environmentalists along the world.

Many of the issues have risen from the recent activities and damages caused by the humans who only are present in the earth for no more than 100,000 years. However, changes in the physical and chemical environment have been experienced from the very beginning of the earth geological history of more than 4,000 millions of years and the issues for living organisms are have been existing since the emergence of the first living beings. Though the relevance of the issues is now relevant to the ones that exist in the current times for humans, it is important to the reader to know that and also to understand that all humans could do actions to help to restore the damaged ecological equilibriums that have been lost.

This book brings the primary focus of the readers to the contemporary issues that concern the environment in current times. It hopes to awaken the minds of the people in that regard and update them on the important issues that surround the survival of people and the existence of the environment.

The salient features of this book are:

- The book introduces both kinds of environment that involve social and physical ones. It throws light upon the environmental crisis that exists and the features it showcases. It discusses the effects that globalization has and explains dual nature it shows. The book also talks about the need for ethics when the environment is in consideration.

- The book discusses the effects of pollution and the threat it poses as an issue, to the environment. It explains the different kinds of pollution that exist in the environment and the things that cause them. It also talks about the impact each kind of pollution has on the environment and the reason that these kinds of pollution persist in certain regions or countries.
- The book gives a high relevance to the subject of global warming and climate change.
- The readers are further informed about the role of an increasing population that is putting unwanted pressure on the environment. The book discusses the way in which the growing population results in decreasing forests. It informs the readers about the overconsumption of resources and the wastage that goes along with it. It also throws light on the other impacts that population has on the environment.
- Going further, the book informs the readers about the negative effect of rising and changing temperatures on the environment. It discusses about the imbalance that exists between the benefits of rising temperatures and the threats it poses on the environment. The book also points towards the observed changes in climate on the global level.
- The readers are informed about the health problems of the humans and their relation with the environmental degradation. The book explains the readers about the air-borne diseases and the way in which they contaminate the environment. It also brings the focus of the readers to the human sanitization and explains its effects on the deterioration of the environment.
- There is further discussion about the role of energy production and its use in the degradation of the environment. The book also explains the various impacts of the use of non-conventional energy resources and the increased pressure on biomass due to its overuse as an energy alternative. There is a description of the impacts that the use of geothermal energy has on the environment.
- The book discusses about the significance of soil in the nurturing of the environment and its preservation. It explains the study of soil science. It also explains the ways in which soil gets contaminated on a regular basis and in turn also put pressure on the environment. It discusses the subject of soil erosion and its implications.
- The book concludes the discussion by talking about the ways in which the people can contribute towards saving the environment. It talks about the unfinished and unaccomplished goals from the millennial development goals set by the United Nations. It points out the importance of proper correspondence between the nations to treat this problem as an international one.

Chapter 1

ENVIRONMENT AND ITS ISSUES IN THE CONTEMPORARY WORLD

LEARNING OBJECTIVES:

In this chapter, you will learn about:

- The basic notion of the environment and its management.
- Critical issues related to the environment.
- Environmental crisis and its features.
- Impact of globalization on the environment.

Key Terms:

- Biodiversity
- Deforestation
- Environment management
- Global warming
- Globalization
- Pesticides
- Pollution
- Stratospheric ozone depletion
- Unprecedented crisis

1.1. INTRODUCTION

In the medical sense, the environment includes the surroundings, conditions, or influences that affect an organism. The environment can also be defined as the sum total of the elements, factors, and conditions in the surroundings which may have an impact on the development, action or survival of an organism or group of organisms. According to this definition, the environment would include anything that is not genetic, although it could be argued that even genes are influenced by the environment in the short or long-term.

Environment and economics go hand in hand. Even though mainstream economists take no account of the environment's role in a country's economy, one cannot deny the fact that economic processes cannot be detached from the natural environment in which they operate. Thus, it is safe to treat the environment as a commodity. This means that economic value can be attached to environmental factors even though this economic value may change with time.

Whatever other value the society may impute to an environmental resource, it will typically also impute an economic value to it. An environmental resource may not have any value to a community, but over a period of time, this value may suddenly increase due to technological changes, or it may decrease owing to change in its utility. However, a common problem in all societies is that the economic value to a person or household of an environmental resource differs, and typically falls short of its value to the community as a whole. Therefore, the stimulus for individuals and households for protecting the resource base is less than what is collectively found desirable.

An economy cannot operate without the constant flow of matter and energy from the environment. However, burgeoning human population and increasing economic activity have given birth to unplanned developmental activities. This has caused enormous strain on environmental resources and thus posing several threats of environmental degradation at local, regional, and global levels. Globally today there are threads like: (1) going to a 6th biodiversity extension, (2) an increase of three degrees in average global temperature, (3) pollution of all sources of water for consumption (Figure 1.1).

Figure 1.1: The major challenge in the current scenario is to save the environment from certain issues that harmful for living organisms.

Source: https://cdn.pixabay.com/photo/2017/04/02/19/23/environment-2196690_960_720.jpg

Therefore, it is imperative to maintain a balance between the capacity of the environment and the sustainable utilization of resources. This can only be possible by developing an in-depth understanding of the environment and the principles of its scientific management.

There has been a steady increase in human population over the last two decades, which has resulted in an increased demand for developmental activities. No development activity is complete without using environmental resources. Since the increasing demands have led to the overuse of natural resources, it is imperative to understand the issues that relate to it and devise mechanisms to deal with these issues.

Environmental management is therefore concerned with the description and monitoring of environmental changes, with predicting future changes and with attempts to maximize human benefit and to minimize environmental degradation due to human activities.

Environment management (EM) is an emerging concept to ensure judicious use of environmental resources. Development activities related to population growth can lead to environmental degradation, increased pollution and depletion of resources.

Thus, EM plays a critical role in maintaining a balance between the capacity of the environment and developmental requirements. This task can be achieved through the application of sustainable development/utilization mechanisms which are based on scientific principles of environment.

Ever since humans started populating the earth and started exploring natural resources, the process of environmental degradation started. Humans have evolved in a big way, graduating from the Stone Age to the age of the Industrial Revolution. Rapid growth in population, expansion of cities, and emerging industrial production intensified environmental problems globally.

These problems are diverse and dramatic, because they encompass practically all constituents of the natural human environment: organic and inorganic nature, natural resources, and climate. Anthropogenic pressure on the Earth's environment reached its peak in the 20th century resulting in a number of irreversible consequences manifesting themselves in the 21st century.

Increased economic activity, irrational use of resources, wasteful practices, overpopulation, urbanization, and armed conflicts has inflicted our planet with damages which can have catastrophic effects. Thus, it is in the larger good of future generations that a proper framework must be implemented with regard to the use of environmental resources.

1.1.1. Physical and Social Environment

Environmental factors, broadly defined, are critical to understanding the health and well-being of the population. Such factors include the resources that individuals have in their physical surroundings, their perceptions of the quality of their neighborhood and communities, and the nature and extent of their personal social networks.

As discussed in the previous section, environmental resources and conditions may have different value to different communities. A detailed study of variables such as gender, age, and occupational class helps in understanding variations in the health and economic status of population subgroups. These factors may affect health and economic outcomes directly and indirectly.

Social is a space that blends the best of the office and the café.

It is important to understand in detail about the range of personal, **social**, economic, and environmental factors that influence the health of older population. Housing conditions are an important aspect of the quality of life. Housing ownership constitutes a major chunk of the wealth of the individual or of the individual's household, and has therefore been the focus of research examining the impact of economic factors on health and mortality. Housing ownership and housing quality are directly linked to mortality and health.

House ownership contributes towards better functioning ability in old age. Owning a house imparts a sense of stability and security in old age, which automatically translates to better health. Recent studies have found that the wealth building effect of homeownership and

Healthy housing is essential for a sustainable and equitable future. While overcrowding results in the spread of illnesses and deprives people from access to resources, under crowding may result in isolation and loneliness for the older generation.

the sense of control it provides to homeowners in a stable housing market affect homeowners' mental and physical health in a positive way. Ownership of durables and other housing assets also contributes positively in better health of older people.

Housing quality is another defining factor for determining the health of the older population in developed countries and urban communities. According to a study conducted in Britain in 2006, poor housing conditions increase the risk of severe ill-health or disability by up to 25% during childhood and early adulthood (Figure 1.2).

Figure 1.2: Overcrowding is a major concern in the contemporary world in context to the environment and its emerging issues.

Source: https://live.staticflickr.com/1436/613445810_2249c2d193_b.jpg

Emerging economies like Brazil, Russia, India, and China, called the BRIC, also face similar challenges. And in all countries, there are always cities were people concentrate and end for being overpopulated.

The vicinity in which people live and the amenities to which they have access can significantly affect the health and well-being of the older population. As compared to the young population, the older population prefers local amenities like grocery and healthcare facilities to be within their reach. These amenities also provide a platform for socializing and meeting people.

On the other hand, the younger generation doesn't mind distant access to amenities. The growth of 'out-of-town' shopping malls has created a new way of shopping. While these malls often provide cheaper options for shopping, only older people with access to good transport can take advantage of these outlets. Another important amenity required by the older population is the banking and financial services.

In spite of technological advancements in financial and banking services, many older people still rely on traditional methods for banking and claiming their benefits.

1.2. THE ENVIRONMENTAL CRISIS

1.2.1. An Unprecedented Crisis

The rapid growth of civilization and industrialization has posed a great threat to the environment. Developmental activities are being carried out at breakneck speed without taking into consideration its effects on the environment. Mankind is staring at an environmental crisis, which many experts claim is unprecedented in its magnitude, pace, and severity. It was not until some of the worst man-made disasters happened that the world started becoming aware of the environmental crisis.

The Sahelian Droughts and Famine struck the Sahel region from West Africa to Ethiopia in 1968 and continued till 1985. As a result of this drought, almost 100,000 people lost their lives due to food shortage and disease (Figure 1.3).

The Bhopal Gas Tragedy in 1984 was another unfortunate incident which left almost 20,000 people dead and contaminated the groundwater of Bhopal city.

Figure 1.3: Environment crisis.
Source: https://cdn.pixabay.com/photo/2017/02/27/08/50/cyclone-2102397_960_720.jpg

The residents of this city suffered from the consequences of this industrial catastrophe for many years. A major assessment of the global environment published in 1999, the UNEP Global Environment Outlook 2000 report, drew attention to two critical, recurring themes:

- The global human ecosystem is threatened by grave imbalances in productivity and in the distribution of goods and services. While some part of the human population is privileged to receive good and services, there is a large proportion of the human population which is deprived of the benefits from economic and technological development and lives in abject poverty.

- While coordinated efforts are being made towards achieving internationally coordinated environmental stewardship, improvements in environmental protection due to new technologies are being outpaced by the magnitude and pace of human population growth, and economic development.

As a result of these factors, a wide range of environmental problems has emerged. These problems include anthropogenic climate change (global warming), depletion of stratospheric ozone, air pollution, acid rain, water pollution, depletion in the water table, deforestation, depletion, and extinction of species, and destruction of **biodiversity**.

Biodiversity is used to describe the immense variety and richness of life on Earth.

All of these problems have serious ramifications on the environment. Even though they manifest in different physical forms, their solution lies intrinsically in human attitudes, beliefs, values, needs, desires, expectations, and behaviors. While these problems do require solutions by environmental 'specialists,' what is more important is to remember what it means to be human.

1.2.2. Main Features of the Environmental Crisis

Accelerated growth in population and economic and development activities has given rise to a grave environmental crisis. This has manifested itself in many forms ranging from exploitation of land to nuclear issues.

It is essential to emphasize that a wide range of views about the nature and severity of the current environmental crisis exists, and some of the issues are highly controversial. Nevertheless, there is broad agreement that the environmental crisis encompasses the following main issues.

- **Climate Change**: It occurs when the Earth's climate system results in a new climate pattern which lasts for at least a few decades. Pollution of the atmosphere by greenhouse gases (GHGs) (and other contaminants) has given rise to anthropogenic climate change. The climate system of the earth receives most of its energy from the sun and some energy from the earth's interior.

The balance of incoming and outgoing energy shapes the earth's climate pattern. If incoming energy is higher than outgoing energy, the earth experiences a warm climate. If outgoing energy is higher than incoming energy, the earth experiences a cold climate (Figure 1.4).

Figure 1.4: Fossil fuel combustion.

Source: https://c.pxhere.com/photos/cb/fd/fire_embers_hot_red_yellow_brand_flame_burn-1158233.jpg!d

Combustion of fossil fuels, emissions from agriculture and industries, vehicular emissions and clearance and burning of forests – all these factors contribute to climate change. The current state of climate change has resulted in global warming – the average temperature on earth has risen by about 0.8°C since 1880. This has resulted in the melting of polar ice caps, an increase in sea level and change in rainfall pattern.

- **Stratospheric Ozone Depletion**: The ozone layer protects the earth from harmful rays emitted by the sun. Ozone present in the stratosphere absorbs most of the ultraviolet (UV) rays released by the sun from passing through the earth's atmosphere.

There has been a significant depletion of the stratospheric ozone due to atmospheric **pollution** triggered by ozone-depleting substances like halocarbon refrigerants, solvents, propellants, and foam-blowing agents (chlorofluorocarbons (CFCs).

Pollution is the introduction of contaminants into the natural environment that cause adverse change.

The lack of protective ozone at high altitudes has resulted in increased levels of harmful ultraviolet radiations reaching the earth's surface, causing a range of health-related and ecological impacts. Therefore, stratospheric ozone depletion has become a matter of serious concern.

- **Degraded Air Quality**: Air pollution is another environmental issue clouding the world today. As per the World Health Organization (WHO), air pollution kills an estimated 7 million people every year (Figure 1.5).

WHO data shows that every 9 out of 10 people breathe air contaminated with pollutants such as particulate matter, tropospheric ozone, oxides of nitrogen, oxides of sulfur, lead, and various aromatic compounds (such as benzene).

Figure 1.5: One of the reasons behind degraded air quality.

Source: https://media.defense.gov/2012/Jun/29/2000138247/780/780/0/120628-F-JQ435-029.JPG

These substances not only degrade the air quality but also cause damage to vegetation. As per data available only in easily available public resources like Wikipedia, at least 140 million people breathe air 10 times or more over the WHO safe limit, and 13 of the world's 20 cities with the highest annual levels of air pollution are in India. Due to the alarming effects of air pollution on the ecological system, it has become a major cause of worry for environmentalists and scientists.

- **Degraded Water Quality**: Water is essential for life to thrive on earth. In the past couple of decades, there has been a severe decline in the water quality. Contamination in water gives rise to a range of health-related and ecological effects (such as the degradation of coral reefs).

Releasing industrial effluents into water

bodies, oil spills, discharging household and agricultural waste into water and accumulation of plastics in water bodies are some of the major sources of water pollution.

Elevated temperatures can also lead to water pollution as it decreases the oxygen levels, which in turn makes water bodies unfavorable for marine life to thrive. Many other human activities, like mining and industrial processes, create toxic effluents, which leads to degradation in water quality.

- **Scarcity of Fresh Water:** 70% of the earth's surface is covered with water, but only 2.5% of the water is available as fresh water. Out of this, only 1% of the freshwater is available for use, the rest being trapped in glaciers and snowfields. Growing human population has mounted the demand for freshwater.

Globally, we consume around 4 trillion cubic meters of fresh water a year. With the limited availability of fresh water, it is becoming a challenging task to meet the increasing demand. Besides the pollution of freshwater sources, poor water resource management practices have resulted in a scarcity of fresh water for drinking.

Irresponsible use of water for irrigation, rampant construction activities, and salinization of agricultural land has deeply affected the water table. The world is heading towards a situation where nations will be scrambling for fresh water. Necessary checks need to be placed in time to avoid such a grave situation.

- **Land Contamination:** It occurs when potentially hazardous chemicals waste or oil is present in a given area of land. Industrial activities often result in chemical or radioactive pollution, which makes land unsuitable for organisms to exist and unsuitable for other use as well. Soil erosion, soil salinization, and soil degradation are some of the effects of land contamination. It has been proven that land contamination may cause profound ecological effects and have important consequences for agricultural and pastoral productivity.

For example, the Guiyu Dump in China is the world's biggest e-waste landfill. A total of 52 square kilometers of land is buried in iPhones, Galaxy S4s, and other famous electronic devices. These devices contain high levels of lead, which leeches into the soil and makes the surrounding land unfit for agricultural use and construction. It is important to understand the catastrophic effects of land contamination and devise mechanisms to reduce it to a minimum.

- **Deforestation:** Forests cover more than 30% of the earth's land. Forests are responsible for producing oxygen, uptake of atmospheric carbon dioxide, and for the existence of living beings. Deforestation has seen a rapid increase in the past few decades, with the most concentrated deforestation occurring in tropical rainforests. It has been estimated that if deforestation continues at this pace, there will be no rainforests within 100 years.

Deforestation occurs for a variety of reasons, agriculture, and pastoralism being the major cause. Other reasons include the destruction of trees for raw material (for paper, charcoal production and timber), construction, and overpopulation. In the Amazonia, the role of intensive mining is impairing large extensions of pristine tropical rainforests as they are initially deforested.

Deforestation is the removal of a forest or stand of trees from land which is then converted to a non-forest use.

While tropical rainforests cover only around 6% of the earth's surface, they are an essential part of the global ecosystem as they help to regulate climate, prevent soil erosion; and are home to a vast number of plant and animal species. Many countries across the world have taken notice of the potential effects of **deforestation** and have taken initiatives to increase forest cover with reforestation. For example, the total forest cover in India is 21.54% of the total area of the country.

- **Land Use Change and Habitat Loss:** Land use change refers to the process in which human activities alter the natural landscape to accommodate changes in population, industry, and other demands. For example, more than 90% of the forest cover was converted to agricultural pastures in England. In Venezuela, the high deforestation of the rain forest in the south of the Orinoco River has changed the color of the previously low sediment rivers given increasing erosion and sediments transport.

Between 2015 and 2017, India added 6,778 square km of forest cover and extended 1, 243 square km of tree cover.

During the last centuries, there has been a considerable amount of construction activity happening around the Bolivian mountains in urban areas. This has posed a great threat to the population as the risk related to environmental degradation increased the potential for landslides as the recently experienced in April 2019 at La Paz, leaving homeless more than 300 people. Issues arising out of change in land use often overlap with issues such as deforestation but have a broader scope.

- **Biodiversity Loss**: Human activities like deforestation, water pollution, and air pollution have threatened the survival of plant and animal species. While species like passenger pigeon and dodo went extinct earlier, species like house sparrow and tigers are facing the danger of extinction. In

1999, UNEP estimated that one-quarter of the world's mammal species and around one-tenth of the world's bird species faced a significant risk of total extinction.

Biodiversity loss is not just confined to terrestrial ecosystems, but serious concerns have been raised about the future of marine and coastal wildlife species as well. The Great Barrier Reef in Australia is a glaring example of how damaging human activity can be to the survival of biodiversity.

In the past three decades, it has lost half its coral cover, pollution has caused deadly starfish outbreaks, and global warming has produced horrific coral bleaching. It is high time we understand the need to protect biodiversity on the green planet and practice restraint in using natural resources.

1.3. GLOBALIZATION AS A "DOUBLE-EDGED SWORD"

Globalization refers to the increasing interdependence amongst nations as a result of cross border trade of goods and services. Globalization benefited the world with opening up economies to trade with other countries, thus increasing the scope of growth of business organizations. The emergence of new global players has blurred the lines between national and international trade and has given rise to a global governance system.

The constantly developing International relations among sovereign states are now insufficient to analyze and understand the political, economic, and social dynamics of the current world. The growing interdependence among states, which cannot be regarded as independent and autonomous entities, has been described with the help of the Billiard Ball Model of World Politics. In this model, the states are billiard balls that collide with one another (Figure 1.6).

Figure 1.6: Globalization and environmental issues.

Source: https://c.pxhere.com/photos/91/d4/protect_ecology_protection_tree_responsibility_globe_earth_world-1334150.jpg!d

Sovereignty acts as protection against the impact of the collision. Since all balls are of different sizes, international politics gives attention to the interests and behavior of powerful nations. Most recently, nations across the world have initiated collective efforts towards tackling problems such as **global warming**, food shortage, terrorism, and the spread of weapons of mass destruction.

Global warming refers to global averages, with the amount of warming varying by region.

This gave rise to the Cobweb Model of World Politics, which states that such a web of relationships has created a condition of complex interdependence, in which states are forced to co-operate with each other.

The widespread use of digital technologies that allow individuals to communicate and the fact that the most prominent issues in the world, as the environmental ones, are inherently transnational, have made raised doubts on the validity of the above-mentioned theories.

Nations are nowadays collaborating not only for economic activities but also for dealing with issues that affect the world as a whole. In the current system, policies are formulated, keeping in mind the interest of all concerned countries regardless of the size of the economy. This has resulted in a world where every nation is treated at par with other nations. While globalization has made international boundaries less distinct, there are still certain areas where sovereignty plays a dominant role.

The failure of the international society in addressing environmental problems such as climate change reflects the need for a reform in the international institutions of the UN system or even the creation of new ones. These institutions should be developed with a global focus and capability to manage and handle situations involving long-term issues. Political and economic ethics based on the ecological system of freedom and justice must regulate their decisions and be down streamed to all signature nations.

Widespread industrialization and commercialization are leading to a scarcity of resources, and the world is facing a possibility of the outbreak of tensions, conflicts, and wars; hence, there is an unprecedented need for cooperation. A wide range of environmental issues are plaguing different economies across the globe, and it is high time we stop taking the environment for granted. It is in the larger good of the society and to ensure a safe future of mankind that nations call global cooperation and coordination in tackling the environmental crisis.

1.4. THE NEED FOR ENVIRONMENTAL ETHICS

Environmental ethics is the discipline of environmental philosophy that advocates extending the traditional boundaries of ethics from human beings to the environment and its nonhuman contents. The basic premise of environmental ethics is practicing empathy towards natural landscapes, resources, species, and non-human organisms. It is a cluster of beliefs, values, and norms regarding how humans should interact with the environment.

Uninhibited human activities have hampered the environment, which in turn can have consequences detrimental to the future of the generations to come. Environmental ethics exerts influence on a large range of disciplines including environmental law, environmental sociology, ecological economics, ecology, and environmental geography. It is therefore imperative to have basic scientific knowledge about the environment in order to formulate policies that resonate with environmental ethics.

The strains of the ecological crisis are so apparent that the task to preserve the environment is the need of the hour. Mankind has been exploiting natural resources since times immemorial, and it is time to strike a balanced relationship between humans and nature. Human beings have relied heavily on the environment to satisfy their needs.

With the increasing deterioration of ecological systems, human beings have realized that economic and judicial methods alone are not sufficient to solve the problems of environmental pollution and ecological imbalances. What is more important is to change our attitude towards nature and to stop taking the environment for granted. Human beings must make concerted efforts to establish a new ethical relationship with nature, wherein there is empathy, love, and respect towards the different elements of the environment.

Ethics act as a guideline for our way of life. They are the standards we employ (among other factors) to determine our actions. Ethics help us in differentiating between the right and the wrong, the good and the bad. They also tell us what we should or ought to do and which values we should or ought to hold. Ethics also help us in identifying how and why we should value certain things and what actions would reflect those values. The definition of ethics is very generic and can vary from individual to individual – an action that may be ethical to one person can be unethical to another person.

Environmental ethics is a new area of study within the larger and older field of ethics. In the early 1970s, a small cadre of philosophers began to realize that underlying our concern for and discussions about land use, biodiversity loss, and pollution were very real, interesting, and new ethical questions rose. We also began to see that complex philosophical notion lay at the core of our disagreements about what we should do with land, how we should value other species, and which policies we should enact to mitigate the pollution.

Environmental ethics has emerged as a philosophical discipline that establishes the moral and ethical relationship of human beings with the environment. Experts suggest that more than policy-level work, inculcating environmental ethics forms the basis of environmental protection. Human beings must develop sensitivity towards the harmful effects of their activities on the environment. It is a moral responsibility for us to protect that earth that has been benefiting us from its resources.

The emergence of environmental ethics has resulted in an increased awareness of the rapidly growing world population and its impact on the environment. People have become privy to the consequences of the growing use of **pesticides**, technology, and industry on the environment. The field of environmental ethics exerts moral responsibility on mankind to conserve and save the environment from dangerous consequences.

Pesticides are substances that are meant to control pests, including weeds.

It helps us in weighing the pros and cons of our actions in light of its repercussions on the future of our planet. Thus, environmental ethics is descriptive in nature and has no specific international environmental code. Rather, it simply tries to outline how humans should relate to their environment, how we should use the natural resources, and how we should treat other living beings.

REVIEW QUESTIONS

1. Define environment.
2. What do you mean by "Environment Management"?
3. Describe some environment-related issues.
4. What are the main features of the environmental crisis?
5. Explain stratospheric ozone depletion.
6. Why environmental ethics is a critical concept?
7. What are the impacts of the environmental crisis?
8. Define globalization.
9. Why globalization is considered to be a "double-edged sword"?
10. What do you mean by "Ecological Imbalance"?

REFERENCES

1. Barman, M., (2017). *Importance of Environmental Ethics and its Approaches in Our Present Society* [eBook] (pp. 116–119). Annual International Conference Proceedings. Retrieved from: http://www.internationalseminar.org/XVIII_AIC/TS4/Mayuri_Barman_115-119_.pdf.
2. Castro, P. J., (2015). *Environmental Issues and International Relations, A New Global (Dis) Order – The Role of International Relations in Promoting a Concerted International System* [eBook] (p. 19). Retrieved from: http://www.scielo.br/pdf/rbpi/v58n1/0034-7329-rbpi-58-01-00191.pdf (Accessed on 18 June 2019).
3. Janevic, M., Gjonça, E., & Hyde, M., (2004). *Physical and Social Environment* [eBook] (p. 16). Institute for Fiscal Studies. Retrieved from: https://www.ifs.org.uk/elsa/report03/ch8.pdf (Accessed on 18 June 2019).
4. *Modern Global Environmental Problems*. Retrieved from: http://planetaryproject.com/global_problems/eco/ (Accessed on 18 June 2019).
5. Nelson, M., (2002). *Introduction to Environmental Ethics* [eBook]. Ethics for a small planet: A communications handbook. Retrieved from: http://www.michaelpnelson.com/publications_files/nelson_esp_intro_chart_2002.pdf (Accessed on 18 June 2019).
6. *Principles of Environment Management*, [eBook] (p. 66). Retrieved from: https://nptel.ac.in/courses/120108004/module1/lecture1.pdf (Accessed on 18 June 2019).
7. *The Environmental Crisis*, (2019). Retrieved from: https://www.soas.ac.uk/cedep-demos/000_p500_esm_k3736-demo/unit1/page_11.htm (Accessed on 18 June 2019).
8. *What is the Environment in the Context of Health?* [eBook] (p. 3). Retrieved from: https://www.who.int/quantifying_ehimpacts/publications/preventingdisease2.pdf (Accessed on 18 June 2019).
9. World Institute for Development Economics Research of the United Nations University, (1992). *The Environment and Emerging Development Issues* [eBook] (p. 33). Retrieved from: http://

dlc.dlib.indiana.edu/dlc/bitstream/handle/10535/7980/the%20environment%20and%20emerging%20development%20issues.pdf?sequence=1 (Accessed on 18 June 2019).

Chapter 2

POLLUTION: A MAJOR ENVIRONMENTAL ISSUE

LEARNING OBJECTIVES:

In this chapter, you will learn about:

- Pollution and its types.
- The science of pollution.
- Major causes of environmental pollution.
- Status of environmental pollution in developing nations.
- Environmental impacts of pollution.

Key Terms:

- Acid rain
- Air pollution
- Carbon monoxide
- Electromagnetic pollution
- Neuromuscular blockage
- Pollution
- Radioactive
- Soil pollution
- Water pollution

2.1. INTRODUCTION

Environmental pollution is defined as the contamination of the physical and biological components of the earth or atmosphere system to such an extent that normal environmental processes are adversely affected. Environmental pollution is one of the most serious problems facing humanity and other life forms on our planet today. Not every pollutant is harmful and endangers human life, but pollutants that are said to be contaminated are those that are in excess of natural levels. Any use of natural resources at a rate higher than nature's capacity to restore itself can result in pollution of air, water, and land.

There has been increasing global concern over the public health impacts attributed to environmental pollution, in particular, the global burden of disease. The World Health Organization (WHO) estimates that about a quarter of the diseases facing mankind today occur due to prolonged exposure to environmental pollution and its increasing levels.

Pollution is the adulteration of the environment by the introduction of contaminants that can cause damage to the environment and harm or discomfort to humans or other living species. It is the accumulation of another form of any substance or form of energy to the environment at a rate faster than the environment that can accommodate it by scattering, cessation, reprocessing, or loading in some meaningless form.

No doubt, environmental pollution is one of the dangerous challenges that the world is facing today. The reason behind its never-ending increasing levels is the industrial revolution. Due to the continuous development of the states, cutting of trees, building construction over productive lands causing irreparable damage to the Mother Earth.

When people discuss the possible causes of environmental pollution, mostly focused on fossil fuel and carbon emissions, but there are different contributing factors also these days. Chemical pollution in bodies of water contributes to illnesses.

Electromagnetic pollution has effects on human health but is uncommonly considered in present times despite the fact we essentially expose ourselves to it on a daily routine and which has become our necessity before and after bedtime. Looking at the causes and effects of environmental pollution will pull any mind on a speedy downhill curved. Solutions are in the works and, if all work together across the world, there is hope remaining, at least for the time being.

Pollution is something that can take many forms such as the air we breathe is polluted, the water we drink is polluted, the soil we use to grow our food is polluted, the lit-up skies also the increasing noise we hear every day can all contribute to health problems and a lower quality of life with major distractions and effects on wildlife and ecosystems.

One of the greatest problems that the world is facing today is that of environmental pollution, which is causing grave and irreparable damage to the natural world and human society with about 40% of deaths worldwide being caused by water, air, and soil pollution and coupled with human overpopulation has contributed to the malnutrition of 3.7 billion people worldwide, making them more vulnerable to diseases.

Pollution is defined as the introduction of harmful substances or products into the environment, which is typically foreign substances, particularly a contaminant or toxin that produces some kind of a negative or harmful impact on the environment or living beings. Pollution presents in many forms ranging from chemicals in the form of gases or liquids, noise, energy sources such as light or heat, or solids such as the types of waste that end up in landfills.

Even naturally occurring substances, such as carbon dioxide and mercury, can be considered polluted when additional quantities are added to the environment in unsafe proportions by industries which can have detrimental effects on the environment and all forms of lives (Figure 2.1).

Figure 2.1: Illustrating the contribution towards environmental pollution.

Source: https://cdn.pixabay.com/photo/2017/06/04/04/16/warming-2370285_960_720.jpg

Pollution happens when the natural environment cannot destroy an element without creating harm or damage to itself. The elements involved are not produced by nature, and the destroying process can vary from a few days to thousands of years. In other words, pollution takes place when nature does not know how to decompose an element that has been brought to it in an unnatural way, such as the use of plastics. Every day the Government trying to restrict the use of plastics and plastic bags and taking many initiatives by imposing fines also to reduce its use, since it is harmful to the environment and cannot be decomposed naturally. Pollution must be taken seriously, as it has a negative effect on natural elements that are an absolute need for life to exist on earth, such as water and air. Indeed, without it, or if they were present on different quantities, animals – including humans – and plants could not survive.

2.2. THE POLLUTION SCIENCE

Environmental pollution is the release of environmental contaminants, generally resulting from human activity. **Carbon monoxide**, sulfur dioxide, and nitrogen oxides (NOx) produced by industry and motor vehicles are common air pollutants. Human actions have contributed enough to pollute the nation by the direct or indirect alteration with the natural resources such as by increasing radiations levels, chemical, and physical constitution and abundance of organisms. Thus, the problem has increased globally, which attracts the attention of all global leaders as well as the citizens to protect its nation by controlling their actions that have an effect on environment purity.

Carbon monoxide is a colorless, odorless gas produced by burning material containing carbon.

There are various pollution sources include chemical plants, coal-fired power plants, oil refineries, nuclear waste disposal activity, incinerators, large animal farms, PVC factories, metals production factories, plastics factories, and other heavy industry. As per the proven studies by scholars, pollutants can cause disease, including cancer, lupus, immune diseases, allergies, and asthma. Also, adverse air quality can kill many organisms, including humans.

There are different forms of pollution affecting the life of humans in many ways, which are as follows:

- Air pollution;
- Water pollution;
- Soil/land pollution;

- Noise pollution;
- Radioactive pollution;
- Thermal pollution.

2.2.1. Air Pollution

Air pollution is a combination of particles and gases that can reach destructive concentrations both outside and indoors. Its effects can range from higher disease risks to rising temperatures. Soot, smoke, pollen, methane, and carbon dioxide are just a few examples of common pollutants.

It is no surprise for anyone that poor quality of the Air affects the longevity of Human life and kills people drastically. If calculated worldwide, almost 4.2 million premature deaths have been identified approximately due to bad outdoor problems as per WHO. Many developing countries have these air issues indoor also up to 3 billion people who cook their meals by burning biomass, kerosene, and coal, still exposed to indoor air pollution.

As per health science, this air pollution has been the reason for many heart diseases, stroke, and respiratory diseases such as asthma. These are the long-term effects of such air pollution but apart from long-term, there exist short-term problems also such as coughing, breathing issues, sneezing, and cold, eyes irritation and headaches.

Nevertheless, many living things emit carbon dioxide when they breathe, and the gas is widely considered to be a pollutant because it is released in large quantities associated with cars, planes, power plants, and other human activities that involve the burning of fossil fuels such as gasoline and natural gas. Carbon dioxide is the most common of the greenhouse gases (GHGs), which trap heat in the atmosphere and contribute to climate change.

Humans have pumped enough carbon dioxide into the atmosphere over the past 150 years to raise its levels higher than they have been for hundreds of thousands of years. Further, the important component which adds to air pollution includes another pollutant associated with climate change that is sulfur dioxide, a component of smog. Sulfur dioxide and closely related chemicals are known primarily as a cause of **acid rain**. But they also reflect light when released in the atmosphere, which keeps sunlight out and creates a cooling effect.

Acid rain is a rain or any other form of precipitation that is unusually acidic, meaning that it has elevated levels of hydrogen ions (low pH).

Many Countries around the globe are facing various forms of air pollution. One of such country is China which is making efforts to clean up the smog-choked skies from years of rapid industrial expansion, partly by shutting or canceling coal-fired power plants. In the U.S., California has been a leader in setting emissions standards aimed at improving air quality, especially in places like famously hazy Los Angeles.

And a variety of efforts aim to bring cleaner cooking options to places where hazardous cook-stoves are prevalent. It is on the people now who have to take a decision on how to clean their own home and protect the quality of the air they breathe in.

People can protect any home by taking appropriate safeguards against indoor air pollution by increasing ventilation, testing for radon gas, using air purifiers; running kitchen and bathroom exhaust fans, and avoiding smoking. When the people start working on home projects like this, it looks for paint and other products low in volatile organic compounds: organizations such as Green Seal, UL (GREENGUARD), and the U.S. Green Building Council can help.

Finally, to fight against the most emerging issue like global warming, a variety of measures need to be taken, such as adding more renewable energy and replacing gasoline-fuelled cars with zero-emissions vehicles such as electric ones (Figure 2.2).

Figure 2.2: Air pollution.

Source: https://cdn.pixabay.com/photo/2017/11/18/01/16/global-warming-2958988_960_720.jpg

Many automobile companies industry wise has already started working on this part by introducing into the market, fuel-free vehicles such as battery charged scooter and car whose fuel consumption is zero and electricity consumption would be few minimal. On a larger scale, governments at all levels are making commitments to limit emissions of carbon dioxide and other GHGs.

People prefer to travel using their own car doesn't matter; only one individual traveling into one car. This gives rise to **air pollution** and traffic.

Air pollution occurs when harmful or excessive quantities of substances including gases, particles, and biological molecules are introduced into Earth›s atmosphere.

2.2.2. Water Pollution

Water pollution refers to those chemicals or other harmful substances that have accumulated into the water to such an extent that it has become dangerous for the human and other living creatures to consume such water. Oceans, lakes, rivers, and other inland waters can naturally clean up a certain amount of pollution by dispersing it harmlessly.

To visualize how this can happens, use the following example: if you poured a cup of black ink into a river, the ink would quickly disappear into the river's much larger volume of clean water. The ink, however, would still be there in the river, but in such a low concentration that you would not be able to see it. At such low levels, the chemicals in ink probably would not present any real problem. However, if you poured gallons of ink into a river every few seconds through a pipe, the river would quickly turn black. The chemicals in ink could very quickly influence the quality of the water. This, in turn, could affect the health of all the plants, animals, and humans whose lives depend on the river (Figure 2.3).

Figure 2.3: Water pollution.

Source: https://cdn.pixabay.com/photo/2014/11/10/07/10/garbage-525125_960_720.jpg

The Earth's surface is covered by over two-thirds of water covering even less than a one third is taken up by land. The ever-increasing issue of the human population is giving pressure on the lands to build more building to cope up with such increased population. As Earth's population continues to grow, people are putting ever-increasing pressure on the planet's water resources. In a sense, our oceans, rivers, and other inland waters are being embraced by human activities not so they take up less room, but so their quality is reduced.

Poorer water quality means **water pollution**. Everyone is aware of that pollution is a human problem because this problem arises very recently by these continuous developments such as the industrial revolution in the history of development. Before the 19th century Industrial Revolution, people lived more in harmony with their immediate environment.

Water pollution is the contamination of water bodies, usually as a result of human activities.

As industrialization has spread around the globe, so the problem of pollution has spread with it. When Earth's population was much smaller, no one believed pollution would ever present a serious problem. It was once popularly believed that the oceans were far too big to pollute. Today, with around seven billion people on the planet, it has become apparent that there are limits. Pollution is one of the signs that humans have exceeded those parameters of normal living life.

Thus, it is understood that to pollute the water, quantities matter significantly. That is how much of a polluting constituent is unconstrained and how big a volume of water it is released into. The smaller the quantity of a toxic chemical is, lower is its concentration, and little would be the impact it creates even if it is spilled into the ocean from a ship. But the same amount of the same chemical can have a much higher impact pumped into a lake or river, where it will become more concentrated as there is less clean water to disperse it.

Although about 70% of the earth is covered with water, it is becoming a scarce resource day by day. As little as merely 2.5% of water resources are fresh water, and out of that, a very small fraction could be availed as uncontaminated drinking water. Therefore eutrophication will lead as a major consequence of population growth to the strain on the supply of fresh water. Considering a report which says that about 15% of the sphere's population survived in "water-stressed" areas in 2016, the share has been expected to touch 50% by the year 2030.

Two-third of the world's population would be surviving with water scarcities by the year 2025, which according to some authors is attributed to population growth and eutrophication as well as other sources of water pollution such as metals. It is important to consider that population growth is higher in those areas of the globe where water is in high demand already, for example, Southwest, Southeast, and Central Asia, Africa, and Oceania. It will have implications for migration as the demand for water will increase over the 21st Century.

Health hazard is source of adverse health effect on a person potentially.

2.2.3. Soil Pollution

Soil pollution can be defined as to refer the presence of toxic chemicals or pollutants in soil, in such higher concentrations which have the risk to pose a **health hazard** to human health or to the ecosystem. In the case of metals which are presents in the soil naturally, no matter whether their levels are high or not, it can still be risky and continue to pose a threat to life if the levels of the contaminants or pollutants present in the soil exceed the levels that should naturally be present otherwise. All systems achieve local equilibriums with their environments; it can be seen in sulfurous waters where some organisms of tolerant species can live.

This is possible as all kind of soils, be it, polluted or unpolluted, contain a variety of compounds which are naturally present. Such contaminants include metals, inorganic ions and salts, and many organic compounds also. These compounds are related to the soil development and edaphic processes associated with long-term soil microbial activity and decomposition of organisms.

Furthermore, various compounds get into the soil from the atmosphere, for instance with precipitation water, as well as by wind activity or other types of soil disturbances, and from surface water bodies and shallow groundwater flowing through the soil. Some microorganisms are also involved in fixing of some ions from the atmospheric pool like nitrogen from nitrogen-fixing bacteria.

When the amounts of soil contaminants exceed natural levels (what is naturally present in various soils), pollution is generated. There are two main causes through which soil pollution is generated: man-made causes and natural causes.

Basically, one can witness the effect of Soil pollution in plants, animals, and humans alike. The level of its effect depends upon the person who is susceptible to soil pollution as soil pollution effects may affect the living beings (plants, animals, humans) in different ways based on their age, general health status, and type of pollutant.

Though, children are usually more vulnerable in contact to the contaminants, because they come in close contact with the soil by playing in the ground; combined with lower thresholds for disease, this triggers higher risks than for adults. Therefore, it is always important to test the soil before allowing kids to play in the yards, especially for people living in a highly industrialized area to save their health.

Soil pollution may cause a variety of health problems such as headaches, nausea, fatigue, skin rash, eye irritation and potentially resulting in more serious conditions like neuromuscular blockage, kidney, and liver damage and various forms of cancer. Thus, taking proactive steps to fight against the same is an important consideration for upcoming actions.

2.2.4. Noise Pollution

Noise pollution, also known as environmental noise or sound pollution, is the propagation of noise with harmful impact on the activity of human or animal life.

Noise pollution can be defined as unwanted and disturbing sound that can have injurious health effects for human health and environmental quality. The problem with noise pollution, though is that it is not only a complex issue but also a highly determined one.

Figure 2.4: Noise pollution.

Source: https://cdn.pixabay.com/photo/2018/08/04/15/34/noise-pollution-3583915_960_720.png

As a matter of concern, this issue has not been given its due recognition as per public nuisance concern, and so managers seem that were prioritizing more dangerous environmental issues (Figure 2.4).

Indeed, the rise in the ownership and use of private transport over the last century has arguably turned the rumble of noise pollution into something of a grumble. Noise pollution affects the human health in many ways:

- The most immediate effect is worsening of mental health. As an example, people who are living too close to airports will probably be more nervous. Continuous noise can create panic episodes in a person and can even increase frustration levels. Also, noise pollution is a big deterrent in focusing the mind to a task. Over time, the mind may just lose its capacity to concentration.
- Another immediate effect of noise pollution is a weakening of the ability to hear things clearly. Doesn't matter short term or long-term prone to noise pollution, it can still cause temporary deafness. No doubt if the noise pollution remains for a long period of time, there is a threat that the person might go permanently deaf.
- Noise pollution also has the effect on the heart. It is observed that the rate at which heart pumps blood increases when there is a constant inducement of noise pollution. This could lead to side-effects like elevated heartbeat frequencies, **palpitations**, breathlessness, and the like, which may even

Palpitations are feelings or sensations that your heart is pounding or racing. They can be felt in your chest, throat, or neck.

culminate into seizures.

- Noise pollution can cause distention in the pupils of the eye, which could interfere in ocular health in the later stages of life.
- Noise pollution is known to increase digestive contractions. This could be the predecessor of chronic gastrointestinal problems.

Noise pollution have a bad effect on sleep and disturb sleep and frequently awakening, or awakening for long periods, which can be very troublemaking. Even if not awakened by the noise, a person's sleep pattern can be significantly disturbed, and a reduced feeling of well-being can result in the next day. Frequent and prolonged sleep disturbances can result in physical, mental, or emotional illness.

2.2.5. Radioactive Pollution

Radioactive materials refer to those materials or elements that emit bad radiation, thus they are not steady, and they get transformed into some other radioactive or non-radioactive materials. The damage that they can cause depends on the radioactive elements and their half time function (the time needed for their concentration to be reduced to half due to radioactive decay processes).

Fundamentally, the greater the half-time, the lowest the effects on human health would be. Radioactive elements with a short and very short half-time pose a serious threat to human health because of their dangerous effects. Most of the radioactive materials have half-lives of hundreds of thousands of years and, once generated, may persevere in the environment for a very long time to go (Figure 2.5).

Figure 2.5: Radioactive pollution.

Source: https://cdn.pixabay.com/photo/2018/01/11/10/59/pollution-3075857_960_720.jpg

Generally, people understand the radiation with reference to bombs and nuclear explosions. No doubt, these are serious sources of high levels radiation (of high energy), there are many other sources as well that are much more common, practically universal, that generate low levels of radiation which are still remain unnoticed so far.

Many people also consider cellular phones as a source of radiation, and thus all the cell phones, cell phone towers, cordless phones, as well as TVs, computers, **microwave** ovens, broadcast antennas, military, and aviation radars, satellites, and wireless internet are all sources of radiation. And so are the common medical X-Rays. Considering this, the picture of radiation pollution significantly expands. From a few explosions and nuclear accidents happening relatively rarely in faraway places, the picture of radiation pollution expands to a complex matrix covering all the Earth and thus involving all of us everywhere.

Microwaves are a form of electromagnetic radiation with wavelengths ranging from about one meter to one millimeter; with frequencies between 300 MHz (1 m) and 300 GHz (1 mm).

2.2.6. Thermal Pollution

Thermal pollution can be defined as a sudden increase or decrease in temperature of a natural body of water, which may be ocean, lake, river or pond by any human influence. This normally occurs when a plant or facility takes in water from a natural resource and puts it back with an altered temperature. Usually, these facilities use it as a cooling method for their machinery or to help better produce their products.

Whenever we discussed about the pollution, the topic of thermal pollution often doesn't come to mind. People will first think of things like carbon emissions, personal pollution, and waste, and a variety of other changing factors (Figure 2.6).

Thermal pollution is the degradation of water quality by any process that changes ambient water temperature.

Figure 2.6: Thermal pollution.

Source: https://cdn.pixabay.com/photo/2019/01/13/12/57/iceland-3930302_960_720.jpg

However, **thermal pollution** is a real and persistent problem in our modern society; it starts when an industry or other human-made organization takes in water from a natural source and either cools it down or heats it up. They then eject that water back into the natural resource; those changes in temperature alter natural processes that can have disastrous effects on local ecosystems and communities.

2.3. CAUSES OF ENVIRONMENTAL POLLUTION

When it was discussed about the possible causes behind the pollution, it was end up summarizing the same, which is all inclusive and not an end list:

- **Pollution Emerging from the Use of Different Vehicles:** Governments in all countries made and issued through special acts guidelines for mandatory compliance upon the individuals using any type of vehicle such as car, truck or two-wheelers to get their pollution control checked and maintained this minimum level including penalty fees for polluting the environment.
- **Emissions from the Use of Fossil Fuel from Power Plants**: which burn coal as fuel contributed heavily, along with vehicles burning fossil fuels, to the production of smog. Smog is the result of fossil fuel ignition combined with sunlight and heat. The result is a toxic gas which now surrounds our once untouched planet. This is known as "ozone smog" and means we have more problems down here than we do in the sky.

- **Water Pollution is a Major Issue:** Many industries dump wastes into rivers, lakes, ponds, and streams in an attempt to hide wastes from Inspectors under the specific countries Environment and Protection Acts. These water sources feed major crops, and food becomes contaminated with a variety of chemicals and bacteria, causing rampant health problems.
- **Trading Activities:** Including the production and exchange of goods and services. Concerning goods, pollution can be caused by packaging such as use of plastic bags and papers or the containers for the packaging, discharge of spare items on to the roads, making of papers by cutting of trees. All these together contribute to pollution. Today plastic pollution has become a very important issue worldwide as it has been demonstrated how plastics ends into the oceans without being decomposed causing severe damages to the ocean fauna.

2.4. ENVIRONMENTAL POLLUTION IS INEVITABLE IN DEVELOPING COUNTRIES

One of the primary reasons why developing countries struggle with air pollution more than developing nations is that developing nations struggle with challenges that developed nations have already solved. Developing nations instead of learning from the examples of developed countries must often consider that choices are difficult choices between the level of pollutants they release and the growth of their economies. The rising of Green economy and Green marketing of products can, however, increase the competitive advantage for those countries that have learned and base their economies considering environmental and ecological impacts. Costa Rica is moving towards carbon neutrality, and there is a clear government's commitment to have a green economy by creating legal and tax structures that prioritize Nature's health. It is an example of how that approach can benefit the life of citizens with global impacts. Costa Rica holds 5% of the world biodiversity while only occupies 0.3% of the terrestrial surface.

Since there is a pressure on developing nations to grow their economies and remain competitive with the already developed nations of the world, multinationals find an arena as that demand encourages governments in developing nations to invest in dirty but cheap **fuel**. Once a nation is sufficiently industrialized, it can begin pursuing alternative forms of energy, but this process of industrialization itself requires massive amounts of energy, creating a spiral of expansion and pollution that is difficult to outflow.

Fuel is any material that can be made to react with other substances so that it releases energy as heat energy or to be used for work.

Environmental pollution is one of the key issues that the world is facing today. Many reports of the scholars and environment departments keep the track of the most polluted cities and most polluted countries. Governments have introduced many initiatives to combat the battle of fighting against pollution problem. Such as cleaning of Rivers, which has brought attention in recent years and a good example of how it is feasible is The River Thames that crosses the city of London in England (Figure 2.7).

Figure 2.7: Say "No" to plastic bags.

Source: https://live.staticflickr.com/1927/31234762448_77d9a8a95d_b.jpg

Use of plastic bags has increasingly been banned by many Countries like Rwanda, Kenya, Taiwan, New Delhi, Australia, France, and they have imposed a number of penalties on using the same. Since it is very difficult to dispose of the plastics which are again decreasing the productivity of the lands.

Developing countries are more prone towards increasing pollution, and they have fewer facilities available to decrease its amount as compare to developed countries. There are only two things no human or any life form can go without, air, and water. Unfortunately, there is a major crisis worldwide related to environmental pollution, the situation being critical in developing countries and impairing water and air.

Even if in some countries pollution levels

aren't alarmingly high there are enough resources and technologies to combat the issue, the fact that countries currently under economic and industrial growth are faced with worrying levels of environmental pollution still impacts global population as the whole state of our planet is affected.

To take care of air pollution, countries which are currently under development should obviously pick hydraulic run of river power plants, wind, and solar energy, even if initial costs are higher. The lengthy lifespan of these energy sources means that the money invested will be recuperated, and seeing how pollution will be reduced, life quality will increase, and proper conditions will be created for economic growth, so the initial investment is worth the sacrifice.

Since at every level of development, the country we are considering must make a choice between two ever-conflicting goals that is either ensuring that the target is reached through the cheapest and fastest means or reducing environmental impact of the actions needed to sustain economic growth and poverty reduction. However, most go for the first choice, as the second choice is considered too costly and timely affair. As the technology for creating the things economically cost high.

People are rushing these days for their output and ignore the quality of the same, which later either affects them or to the environment. That is why, responsibility is cast on the companies to fulfill their social obligations, and environment protections also come under this. There is a saying, if you cut one tree, seed two trees in another area as a cost of compensation to the environment and its sustainability. However, it cannot really be compensating if the tree species are unique or the trees are so old with low growth rate that will not be feasible to restore them in the forest ecosystem resulting in a depauperate forest with the time. As the path towards economic growth and proper development is crucial and must be followed, besides, it shouldn't affect and result into sacrificing environmental safety and our health.

Though many countries also help in some of the projects, however, relying on the help of others is not the sole solution, leaders being able to fix issues on their own as long as they put the same amount of interest in the manner in which development occurs, not solely focusing on speed and costs. Efficiency and effectiveness together bring the results; one without another would be no benefit for the overall success. Thus, if every corporate and every individual start taking responsibility of their own destructions, this problem would have never arisen. Ecological and economic ethics is needed in all governance structures.

Hence a considerable budget should go towards water treatment plants that tend to the purification of water sources before they reach people's homes. As the water crisis is at an all-time high worldwide, this issue exists even in developed countries; therefore it's all the more imperious for every country to try and fix the situation in its area before increasing large amounts of chemicals are needed to be destined for sanitation. While environmental pollution is inevitable in developing countries, the situation can be compensated at least if the right steps are followed by governments and people. A poor developmental situation on a country impacts the rest of the world as well, so it's time we all take measures to help those in need for their sake, as well as ours. It is projected that over the next decade most of the growth that will occur in developed countries will happen in urban areas within the poorest countries of the world, with many of the countries located in Asia.

Both India and China are seeing periods of rapid expansion, and much of this expansion releases toxic air pollutants that can harm people's health and lifestyles. The high levels of air pollution are thought to cause millions of deaths every year in these countries and millions for hospitalizations and missed workdays. Citizens in these developing nations face a high risk of heart disease, lung cancer, asthma, and stroke when compared with their counterparts' citizens in developed nations.

It should, thus, be also noted that the effect of pollution growth in these countries is not limited to those countries. Prevailing winds and climate events can carry these pollutants thousands of miles to other parts of the world. California and even Alaska are showing the effects of pollutants generated in China. It is now known how the Sahara fine sands arrive to the Amazon forests through the winds. The problems associated herein are a global concern that has to be handled and taken care of on a serious note as it may affect the countries around the globe.

2.5. ENVIRONMENTAL IMPACTS OF POLLUTION AND CHANGES IN ATTITUDE

Natural environment encompasses all living and non-living things occurring naturally, meaning in this case not artificial.

Pollution has affected the **natural environment** so badly that the weather has been changed. Those months which used to be hot and cold every year are now unpredictable. The spring season has shifted, and sometimes rain happened without weather. Even the forecasting of the weather department shows the wrong forecast. In tropical regions of northern South America, unpredictable rains and dry seasons lead to environmental disasters which are exacerbated during El Nino and La Nina years.

Humans can take various steps to save the environment from various pollution sources at individual level, with changes in actions, small actions in millions of people can be more effective than all governmental regulations. For instance simple people actions can help to save the environment from Air pollution: conserve energy no matter where people work or at home, preferring to buy the Air conditioner considering the Energy star level to save consumption, using public transportation or using or offering carpooling for commutation and sharing of expenses, keeping cars and other vehicles always pollution friendly, using environmentally safe paints and cleaning products, which are easy to find nowadays.

To save the environment from water pollution, people can start from home their selves. Save water and avoid its wastage. While people are using the aqua guard at home or RO, always place one article on the lower side where water is drained out by the system for cleaning.

This water can be reused for any other purpose than drinking. On an individual and family-scale, it is possible to make use of some of the same principles. Home water-testing kits are available to assess your drinking water, whether it comes from a public reservoir or a well.

Be sure to report anomalies in the area of metals and ions such as chlorine. Be careful to not waste water through frivolous means, such as leaving sprinklers running when rain is anticipated or has very recently fallen. A smaller burden on municipal treatment and sewage-disposal systems is vital for ensuring the availability of clean water – an indispensable element of good health – for everyone.

To prevent soil pollution, people should stop throwing waste material on the land, always throw in dustbin provided or place one dustbin if not provided. To improve the quality of the soil, plant trees. If the soil is in the playground where kids play, people should try to get it check as whether its pollutant levels are safe for them as kids are more prone to health risk faster.

Researchers in the US have studied the impacts of the air pollution on the human health and provided with the following steps that would help reduce the exposure to dirty air:

- **Filter Air at Home:** Evidence for the effectiveness of air purifiers in reducing the health effects of air pollution is mixed. But they may be helpful for people with allergies. And a small study found that a small group of college students in Shanghai, China, who ran an air purifier in their dorm, had lower levels of stress hormones like Cortisol, blood pressure, and insulin resistance after nine days than those who used a sham purifier. Consumer Reports tests how well portable air purifiers remove dust, pollen, and smoke from the air. Check the air purifier ratings.

- **Ban Smoking and Reduce Fireplace Use:** The smoking of tobacco is a key source of air pollution indoors, and that cozy fire in fireplaces also releases plenty of fine particles. Every city should have the Environmental Protection Agency's advice on burning wood at home. And citizens should be aware of other sources of indoor pollution, such as gas cook stoves.
- **Avoid Particle Pollution:** Combustion from car and truck engines as well as from power plants is the most common source of harmful particles in the air in the U.S. More than half of all air pollution comes from mobile sources, primarily automobiles, according to the EPA.

So, if people exercise outside – and especially if having a heart or lung problem – people should try to avoid heavily trafficked roads where automobile-generated pollution could be significant. It can also check this EPA website for air-quality alerts and, depending on your personal health, consider limiting outdoor activity when the air quality index is 101 or higher.

- **If People Live Near a Busy Road:** may want to keep the windows closed during the day or times of more traffic, such as rush hour. Instead, open them at night, when traffic may be lighter. It makes people less vulnerable. Because the health effects of air pollution may weigh more heavily on those already at risk for heart disease, people should take smart lifestyle steps.
- **Reducing People's Pollution Footprint:** Because so much of air pollution is created by traffic, it has being suggested to consider alternate ways of getting around. "Walk, bike, public transportation," "If there is a need for new car, there are some countries where now is possible to consider an electric or hybrid car." (Check our electric and hybrid vehicle ratings.)

Thus, through this research, it is possible to identify that what can be the possible impacts of increasing pollution and how we can contribute to create a better future and sustainable environment. Governments issue many guidelines towards fighting against the pollution, and people needs to adhere to such guidelines seriously. On the other hand, Governments can use National Pollution Index NPI data to assist with environmental planning and management. NPI data is often used in the preparation of State of the Environment reports, and to support initiatives which help protect the environment. Pollution control is the responsibility of state and territory environment agencies.

REVIEW QUESTIONS

1. Define pollution.
2. What are different types of pollution?
3. Explain pollution science.
4. Describe various causes of environmental pollution.
5. Why environmental pollution is unavoidable in the developing countries?
6. What is "Environmental Kuznets Curve"?
7. Illustrate the role of industrialization in context to the environment.
8. Briefly describe the environmental impacts of different kinds of pollution.
9. Why air pollution is considered to be more threatening kind of pollution?
10. Describe the root cause of radioactive pollution?

REFERENCES

1. Anand, S., (2013). *Global Environmental Issues* [eBook] (p. 9). Open access scientific reports. Retrieved from: https://www.omicsonline.org/scientific-reports/2157-7617-SR-632.pdf (Accessed on 18 June 2019).
2. *Causes and Effects of Environmental Pollution*. Retrieved from: https://www.conserve-energy-future.com/causes-and-effects-of-environmental-pollution.php (Accessed on 18 June 2019).
3. Climate Change Conferences 2019 | Environmental Sciences Meetings | Medical Entomology Congress | Global Warming Symposiums| Europe | USA| Middle East | Asia| UK | 2019, (2019). Retrieved from: https://climatechange.earthscienceconferences.com/events-list/thermal-pollution (Accessed on 18 June 2019).
4. Coker, A. *Environmental Pollution: Types, Causes, Impacts and Management for the Health and Socio-Economic Well-Being of Nigeria* [eBook] (p. 23). Retrieved from: https://pdfs.semanticscholar.org/8e7b/a9595bab30d7ea87715533353c53f7452811.pdf (Accessed on 18 June 2019).
5. *Environmental Pollution – an Overview | Science Direct Topics*, (2017). Retrieved from: https://www.sciencedirect.com/topics/earth-and-planetary-sciences/environmental-pollution (Accessed on 18 June 2019).
6. Gustave, S. J. (1988) *Environmental Pollution: A Long-Term Perspective* [eBook]. World Resources Institute. Retrieved from: http://pdf.wri.org/environmentalpollution_bw.pdf (Accessed on 18 June 2019).
7. Madaan, S. *Primary Causes of Thermal Pollution*. Retrieved from: https://www.eartheclipse.com/pollution/primary-causes-of-thermal-pollution.html (Accessed on 18 June 2019).
8. Omoju, O., (2014). *Environmental Pollution is Inevitable in Developing Countries*. Retrieved from: https://breakingenergy.com/2014/09/23/environmental-pollution-is-inevitable-in-developing-countries/ (Accessed on 18 June 2019).
9. *Pollution*, (2013). Retrieved from: http://www.everythingconnects.org/pollution.html (Accessed on 18 June 2019).
10. *Radioactive Pollution: Causes, Effects and Solutions*. Retrieved

from: https://www.conserve-energy-future.com/radioactive-pollution-causes-effects-solutions.php (Accessed on 18 June 2019).

Chapter 3

GLOBAL WARMING: FROM CLIMATE CHANGE TO THE GREAT ENVIRONMENTAL ISSUE

LEARNING OBJECTIVES:

In this chapter, you will learn about:

- Global warming and its causes.
- Tradeoffs between possible benefits claimed for global warming.
- Critical effects of rise in sea level.
- Changes in the local, regional, and global climate.
- Impact of global warming and climate change on society.
- Diseases caused by the effect of global warming.

Key Terms:

- Anthropogenic climatic alteration
- Climate change
- Dengue
- Desertification
- Ecosystem
- Farming
- Flood
- Glaciers melt
- Global warming
- Malaria
- Phytoplankton

3.1. INTRODUCTION

Warming up on a global level and changes in climate have such a deeper impact on the ecosystem that they can even transform the whole biological system. Further precisely, variations in the temperature of air cover which is near the surface of the earth will possibly affect the functioning of the ecosystem and consequently the biodiversity of animals, plants, and such other sorts of a living being.

Due to acclimatization to long-term cyclical climatic trends, the present physical sorts of the animal, as well as plant varieties, have been ascertained. As warming up of atmosphere changes, such patterns on span of time noticeably smaller than the ones which rose in the ancient times from variability in the biological environment, comparatively abrupt biological alterations might challenge the usual adaptive capability of numerous breeds.

Figure 3.1: Car fuel and emission of harmful gases is a big reason behind global warming.

Source: https://c.pxhere.com/photos/d6/0d/car_non_panne_bad_luck_uitaatgassen_pollution_air_pollution_car_breakdown-618814.jpg!d

Sea levels are rising, glaciers are melting, cloud woodlands are vanishing, and wild animals are struggling to go on. As we power our modern lives, one thing is very clear human being is responsible for the maximum of the present-day global warming due to the emission of heat-trapping smokes. The level of emission of greenhouse gases (GHGs) has broken all the past records of the previous eight hundred thousand years.

Consequences of global warming vary from place to place. It has caused an array of transformation in the climate of Earth, and even variations in the long-term environment patterns. Though most of the times climate change and global warming are used interchangeably, according to scientists "climate change" describes the intricate changes currently upsetting the earth's climate systems as well as weather – in part since specific regions in the short term get cooler actually (Figure 3.1).

Climate alteration includes not merely increasing normal hotness but also shifting wildlife populations and habitats, extreme weather events, rising seas, and many other such side effects. All these variations are evolving because of the reason that human beings carry on to increase heat-trapping GHGs to the air resultantly altering the paces of the environment which all forms of life have to suffer without any other option.

Greenhouse effect is the process by which radiation from a planet's atmosphere warms the planet's surface to a temperature above what it would be without this atmosphere.

To understand what **greenhouse effect** is in a simple way, it is needed first to understand that when sunrays touch the surface of the earth, the solar energy is captivated and then in the form of heat, the same energy is radiated back into the atmosphere. In terms of the greenhouse, molecules of GHG captivate some of the heat, and the leftover gets escaped out of the earth's atmosphere. The more the GHGs concentrated in the air, the more heat is trapped by the molecules. Since 1824, researchers are aware of the greenhouse effect, at that time it was calculated that the temperature of Earth would be much cooler if there wasn't any atmosphere. The Earth's climate is kept livable by this natural greenhouse effect. In case if there would not have been an atmosphere, the surface of the Earth would have been an average of around 33°C (60 degrees Fahrenheit) cooler.

3.1.1. Variation in the Temperature and Global Warming

Earth's climate is not affected by human activity alone. Some other factors that contribute to it could be listed as: variations in solar radiation from sunspots and volcanic eruptions, the Earth's position relative to the sun and solar wind, etc. Similarly, the large-scale climate patterns like El Niño (Figure 3.2).

Figure 3.2: Climate change and global warming are highly inter-related and are impacting the atmosphere adversely.

Source: https://cdn.pixabay.com/photo/2016/12/15/09/07/climate-change-1908381_960_720.png

However, the models on climate which are used by scientists to observe Earth's hotness take all such elements under consideration. Minute particles suspended in the atmosphere from volcanic eruptions as well as changes in solar radiation levels, for instance, have added to merely around 2% of the current global warming. Rest occurs due to human-caused factors, such as land use change and of course, the GHGs.

The short time scale of this current warming is important as well. For example, the Earth's surface gets temporarily cold due to emissions from volcanic eruptions. However, such effects are temporary, and they last for a very few years. Another example is El Niño and La Niña – such events too work on predictable and short cycles. Alternatively, the sorts of universal hotness variations which have added to ice ages happen on a span of multiple lakhs of years.

Since the beginning of life on earth, releases of GHGs into the environment has been counterbalanced by GHGs which are absorbed naturally. Consequently, the concentrations of GHG and temperatures on the Earth had always been quite steady. This is the main factor that mankind civilization could prosper in a steady environment.

A quick increase in GHGs has become a challenge as it is altering the environment sooner than many forms of life can acclimatize themselves to it. Likewise, new and rather more irregular environment stances exclusive problems to all forms of life.

Climate is defined as an area›s long-term weather patterns.

In ancient times, temperatures of Earth were cold enough to cover much of North America and Europe with ice. So, we have seen a regular change in the **climate** of the earth that we see today and what it used to be in the ice age. Coming down to numbers, the variance amid average global temperatures during those ice ages and today is merely 5°C (9 degrees Fahrenheit). The important thing to be noticed here is that this shift in temperature occurred at a very gradual pace, i.e., over a time span of hundreds of thousands of years.

However, with the intensity of GHGs increasing, remaining ice sheets of Earth's as like Antarctica and Greenland have started to liquefy. This additional water will increase sea levels considerably, and rapidly. By the year 2050, as glaciers melt, sea levels are forecast to increase amid 1 to 2.3 feet.

With the rise in temperatures, there may be unexpected changes in the climate. Another matter of concern is that weather can become more extreme, in addition to sea levels rising. It implies further forceful major storms, higher intensity of rain trailed by drier as well as longer droughts – a huge hurdle for cultivating crops – alterations in the ranges in which animals and plants could survive, and shortage of supply of water which have traditionally originated from glaciers.

3.2. TRADE-OFFS BETWEEN POSSIBLE BENEFITS CLAIMED FOR GLOBAL WARMING AND WHEN COMPARED TO THE KNOWN HARMFUL EFFECTS

Many intellects claim that there could also be some advantages of global warming. These people are reluctant to start the programs to discourse the destructive results of environmental alteration. As a certain displacement of present doings could be unavoidable in any big initiative for alteration, such probable advantages must be compared as well as examined to the destructive effects of failing to start operative relaxing events.

In operation, an approximate benefit to cost analysis turn out to be significant when the investment of resources is concerned, which is surely the situation now.

3.2.1. Probable Advantages to Farming at Higher Latitudes

One of the most common foreseen advantages of warming up of atmosphere at the global level is that it has contributed to the lengthier growing season which shall be enjoyed by higher latitude expanses of Asia as well as North America. This movement of high temperature in the direction of north in the northern hemisphere has been well observed and scripted.

But the advantages of global warming could only be realized if other conditions that are conducive to successful agriculture like rainfall to move in the direction of north trailing the high temperatures. However, as of now, nothing can be said about it. Several present general circulation models (GCMs) related to worldwide temperatures have agreed that warming up of global **atmosphere** may make dissimilar alterations in diverse geographic expanses.

Atmosphere is more likely to be retained if the gravity it is subject to is high and the temperature of the atmosphere is low.

Further researches are required prior to making forecasts about the detailed alterations in various geographical locations confidently. The events which validate the delicateness of supposedly rich lands anticipating progress, recent history has two evident examples. One is previous Soviet Premier Nikita Khrushchev's "virgin lands" program for altering central Asia into a massive breadbasket. The attempt was a big-time flop. The other is the American Rocky Mountain West, continually endangered with shortages of water and evidently unable to support huge inhabitants despite of high nineteenth-century publicity on the contrary.

Additionally, a majority of displacement of habitat conditions could be anticipated as new animals as well as plants shift towards new states and out of others. However, it cannot be said that final outcome of the procedure is going to be dangerous on the edge, but it is clearly evident the whole phase of transformation is going to be stressful for all the ecosystems involved, especially when the transformation is quick. Rapid alteration is really improbable to be advantageous from the viewpoint of the health of ecosystem.

3.2.2. Negative Effects of Rise in Sea Level

We requisite to inspect the side effects of the

climatic alteration that fallouts from global warming, against these possible but still questionable benefits. As already discussed, no one disagrees to the fact that there will be many problems due to global warming, out of which the utmost intense would be the increase in the mean sea level.

People residing in the Washington, DC area, it is totally recreational to drop in the exhibition hall of the National Academy of Sciences (NAS) on 6th and E Streets, NW. A practical demonstration exposes the amount of coastal Maryland would be submerged in the waters of Chesapeake Bay when the average sea level rises by 0.5 meters, by 1.0 meter, and by 1.5 meters. An increase of 0.5 meters in sea level is nearly equal to 1.6 feet (Figure 3.3).

Figure 3.3: Rise in sea level.

Source: https://live.staticflickr.com/2702/4266670427_1845095d6a_b.jpg

To mention a reference, the computed increase in the average worldwide sea level, which has previously happened in the 20th century is of the order of 1.5 foot.

When the water temperature rises, the surface water of the ocean at present temperature increases, so due to warming up at global level coastal flooding at several places gets unavoidable. This causes extremely terrible outcomes at the areas that are densely populated and situated nearby sea level. Liquefying of icecaps and glaciers further quickens the procedure.

3.2.3. The Specter of Catastrophic Climate Change

From the impact of global warming on the world's oceans, there could be some other potentially disastrous problem which may occur. To go deeper into the matter, there could be a probable alteration in the worldwide oceanic conveyor belt that manages the coupled deep-water flows as well as several surface currents which has proven significant influences on main local temperatures.

One of the best-studied and best known of such examples is the Gulf Stream. It is famous for playing a vital character in retaining moderate environments within Northern Europe. Not all climate scientists agree on how significant this warm current is to maintain the temperate environments in Northern Europe as well as on how much, if any, reduction in Gulf Stream flow has already begun.

Though, a specific decrease in this flow – known as the thermohaline circulation from the driving force – is unavoidable on the condition that sufficient lower-density, less-salty, meltwater is initiated in the North Atlantic. The same thing is expected to happen from quick liquefying of the Greenland ice sheet, augmented further by liquefying of the Arctic sea ice. So, it has been accepted collectively that some amount of drop in temperatures in Northern Europe is totally unavoidable.

3.3. WITNESSED ALTERATIONS IN GLOBAL, REGIONAL, AND LOCAL CLIMATE

As per a report published early in the year 1989, 1988 was the year that witnessed the warmest climate during the past century over mean world-wide temperatures (the University of East Anglia and British Meteorological Office). Those scientists also stated that 1989 was the 5th warmest year in the past 134 years. For all this analysis they had collected world-wide temperatures histories, and they also found that six out of the ten warmest years on records were in the decade of 1980s.

Though environmentalists argue on whether the witnessed intensification in the previous hundred years of nearly 0.6°C in mean yearly worldwide surface temperatures is a visible indication of warming due to greenhouse gases or yet in the natural inconsistency of weather.

There appears slight disagreement amid scientists which mankind has by now considerably restructured the climate of earth over these elements as the intensification of desertification through **land degradation**, urban heat-island effect, stratospheric ozone depletion as well as deforestation.

Such elements, specific of which have ascended self-reliantly of the accumulation in the worldwide intensification of Green House Gases, are probable to leave substantial impacts on humankind, frequently amplifying the disturbance produced through a quick greenhouse-induced warming up at global level.

3.3.1. Potential Intensification of the Urban Heat-Island Effect

Land degradation is a process in which the value of the biophysical environment is affected by a combination of human-induced processes acting upon the land.

Construction of buildings, roads, pavement, removal of vegetation, and many other mankind alterations of the regular atmosphere, combined with the nonstop generation of heat by social activities, are said to source the temperature of metropolitan areas to increase more than those of neighboring countryside regions.

In the USA such city heat-island effect was assessed at a mean of merely over 1.1°C for a trial of thirty USA metropolises and around 2.9°C for New York City. In USSR, Moscow, the heat-island effect is expected to contribute around 3° to 3.5°C to mean yearly temperature.

The city heat-island outcome in Shanghai, China, is noticeable, with a possible concentration as high as 6.5°C on a peaceful and perfect December night-time in the year 1979, however, no heat-island outcome detected on days with heavy rain as well as strong wind. The city heat-island outcome has been decreased considerably in one of the cities by large-scale plantation of trees. In Nanking, China, the plantation from the year 1949 of thirty-four million trees has been accredited with a noteworthy drop in the mean temperatures of the city.

Though the heat-island outcome has inclined to pass into world-wide environment debates mainly in the background of whether specific of the detected world-wide temperatures increase might be credited to the settlement of specific thermometers in city regions, the heat-island outcome has substantial self-governing importance for humankind.

The quick urbanization of several emerging nations is altering the setting of numerous metropolises. The on-going volatile development of cities such as São Paulo (Brazil), Cairo (Egypt), Lagos (Nigeria), Mexico City (Mexico), Delhi (India) and various other such modernized metropolises is possible to affect a deep variation in the native environment.

It is conceivable that substantial temperatures intensification in various such regions, attributed to the heat-island outcome only, could be added to an already substantial anticipated temperatures rise from **greenhouse** effect induced warming up at global level. Human discomfort, heat-stress impacts on health, and worsening of city air pollution complications, like smog, are some of the likely concerns of an increasing heat-island and greenhouse warming.

Greenhouse is a building where plants such as flowers and vegetables are grown.

3.3.2. The Expansion of Desertification Following Land Degradation

Removal of tree cover, overgrazing of agricultural lands and exhaustive agrarian performs are all human actions that have promoted soil erosion and have fast-tracked desertification. In the past, these things largely happened due to regular climatic procedures. It may hold true or not the present desertification is considerably related to a greenhouse-induced warming pattern, this has caused vast displacements and excessive social distress in Africa.

Thousands of wanderers in Mauritania have been moved into city regions and refugee camps, forming excessive societal chaos. Lt Col Christine Debrah (Ret'd), Executive Chairman of the Environmental Protection Council of Ghana recently summarized the extents of this difficulty as:

We all have gathered today; thousands of Africans are agonizing from malnutrition and hunger. Massive numbers of persons are shifting in and across nationwide borders in hunt of food, thus forming environmental expatriates.

Of course, there is not any exact calculation of the degree of the disaster, however, the events already occurring are indicative of magnitude: an estimated 4 million refugees and returnees and 150 million persons endangered by malnutrition and starvation and an uncountable figure of emigrant people.

In Africa, dry weather environments have been suffered, giving rise to brush-fires and more deterioration of the ground. Certain environment models propose that this tendency in most of Africa is possible to get aggravated as the sphere warms up, through augmented drought conditions in the areas like Northwestern Africa. In South America, increasing aridity is expected from the analysis of results of several of the IPCC models.

3.3.3. Regional Climatic Implications of Deforestation

Large-scale deforestation has not just contributed to the greenhouse effect by transforming trees to carbon dioxide as well as decreasing plants availability to stock carbon dioxide, likewise intensely changes regional as well as local environments. This change may take numerous forms.

Extensive deforestation seems to have been a big element in mainland runoff. Through eradicating vegetation shield, the water retention capacity of the soil is reduced due to deforestation, thus escalating the soil erosion and leaving low lying regions more susceptible to drowning.

In addition, large-scale deforestation seems to dehydrate the environment of the neighboring area by researches proposing these effects in peninsular Malaysia, parts of India, Ivory Coast, parts of the Philippines, and the Panama Canal area and possibly even in Northwestern Costa Rica, southwestern China, and northern Tanzania.

The regional as well as local environmental disturbance, particularly drying of agrarian farmlands, could be of ample bigger environmental influence to several regions than the alterations caused due to warming induced by the greenhouse. Other social actions additional to deforestation may also have regional climatic implications, e.g., summer fallowing, drainage of wetlands, grazing of livestock as well as bush clearing.

3.3.4. Trends in Stratospheric Ozone Depletion

In the past 20 years, there has been a specific depletion of stratospheric ozone above the mid-latitudes of the northern hemisphere. Since the late 1960s, the reduction in entire column ozone amid latitudes 30° North and 64° North have been 3–5.5% in the early spring/winter months. NASA, WMO Ozone Trends Panel; UNEP reported in the equatorial region, and southern hemisphere (apart from Antarctica where there has been noticeable ozone damage) are very scant to be assessed at this point of time.

In spite of the definite ozone loss, the statistics on the rise till now in naturally firm outcomes of UV-B radiation have been nominal. Most of the influences of the detected ozone decrease can happen in the winter season because, at a low angle of sunlight, UV-B radiation is usually low? Therefore, the witnessed escalations have been numerically very minor.

Even that after the complete implementation of the Montreal Protocol to guard the ozone cover, this has been estimated that intensities of chlorine in the stratosphere could increase approximately three folds from their present levels of around 2.7 parts per billion. Additionally, it has been observed that in particular circumstances, substantial ozone loss could happen over assorted reactions in zones such as the Antarctic.

UV-B radiations in the physically active spectrum are proposed to upsurge by at most 20 to 25% by the year 2050, but the proposed reduction may fluctuate by latitude? A rise of wherever nearby this magnitude may have severe concerns on mankind health, agriculture, forests, the maritime food chain, and supplies.

Augmented UV-B radiations resultant from stratospheric ozone reduction will intermingle with warming up at global level to influence the issues like human health, air pollution, fisheries, vegetation, and natural systems. Additionally, it is quite possible that anticipated additional decrease of stratospheric ozone may have a certain straight influence on the temperatures itself by changing the environmental chemistry in the troposphere.

Though this is not at all astonishing that a substantial drip in **stratospheric ozone** may disturb temperatures of the surface, even at this point it is almost impossible to ascertain the consequences of this influence: it may reduce or magnify the proposed degree of global warming. However, augmented UV-B radiations touching the ocean surface may turn to increase the warming by decreasing the biomass of carbon dioxide captivating maritime phytoplankton.

This thing is beyond doubt that connection amid stratospheric ozone depletion as well as global warming is that in both processes other ozone-depleting substances (ODS), as well as the man-made chlorofluorocarbons (CFCs), play a common role. Such substances could deliver as big as 25% of the existing forcing in the direction of global warming and are the main reason for stratospheric ozone reduction.

Stratospheric ozone is a naturally-occurring gas that filters the sun's ultraviolet (UV) radiation.

3.4. INFLUENCES ON OUR SOCIETY AND ENVIRONMENT

The proposed climatic alterations because of anthropogenic effect on the climate of Earth, as it has been noticed, comprise warmer temperature of surface, more widespread drought, shifting patterns of rainfall, increasing levels of sea, as well as further thrilling climatological situations containing intense tropical cyclones and flooding. Such alterations in environment are expected to leave intense influences on human health, **ecosystems**, agriculture, water resources, as well as the elementary arrangement which can support current civilization.

Ecosystem is a community of living organisms in conjunction with the nonliving components of their environment, interacting as a system.

All things considered, influences are probable to be destructive, rather than advantageous, and include widespread extinction of animal species, a superior inclination for loss of water resources and drought in some areas, loss of coastline and coastal wetlands, decreases in global food production, flooding in many regions, increased severity of storm damage, as well as augmented extent of communicable illness. Such extra strains on civilization might, in sequence, give way to bigger struggle.

The degree of all these influences will be subject to the amount as well as rate of warming in future, and on the kind of alterations that have been carried out. In the following section, each of the various impacts has been discussed in more detail.

3.4.1. Ecosystems

Climatic alteration is probable to impact biodiversity as well as to influence the functioning of ecosystems. Animals, as well as plants, have formed their existing terrestrial ranges over long-term acclimatization to cyclical climatic trends. Anthropogenic climatic alteration is probable to change these cyclical trends on a span quicker than have happened certainly over previous ages.

This quick speed of climatic alteration that is probable to test the regular adaptive capability of living beings. For an added warming between 1.5 and 2.5°C, the range of warming foreseen by the year 2100 in even the lower-range releases situations it has been projected that somewhere between a 5th to a 3rd of total animal and plant breeds are expected to face an amplified jeopardy of disappearance.

As proposed by several versions for the higher end releases circumstances (for example, in the above-mentioned A1B setup), for warming in excess of 4.5°C, as high as 40% of breeds would be at jeopardy of disappearance. These extensive disappearances will, in order, probable to intimidate the fine equilibrium as well as tropic arrangements of ecologies, presenting a broader danger to bionetwork services and function and biodiversity.

Climatic alteration is probable to source the devastation of fragile and rare territories which are habitat to "specialized species" incapable to flourish in any external environs. Some examples are: the salt marshes, coastal wetlands, and swamps of mangrove trees which home several unusual breeds are in a state of risk of extinction due to rise in levels of sea.

Particular amphibian classes as like the golden toad whose habitations are confined to specific remote tropical cloud forests, has either extinct or on the verge of extinction because of the reason that such surroundings are disappearing in reaction to global warming.

3.4.2. Farming and Agriculture

The agrarian output may upsurge discreetly largely in extratropical areas in reaction to a limited warming of 1–3°C. Alike upsurges are anticipated for farm animals' production, as it depends on feedstuff s that are themselves agrarian produces. Such intensifications are chiefly a consequence of the expansion of cultivating periods in moderate areas related to large scale warming.

Though, for further bigger quantities of warming, agrarian output initiates to drop down, as this warming initiate to increase beyond the optimum temperatures level for photosynthetic action which plants have developed over long-term progression. In subtropical and tropical areas, crop output is proposed to reduce even for feeble confined warming for the reason that fundamentally any warming surpasses these optimum temperatures level.

The augmented occurrence of flood events and droughts could result in more deterioration in farming and agricultural output, with the influences inducing subsistence agriculturalists utmost harshly. Due to a reduced growing season resulting from warmer and drier conditions, agricultural productivity has already been observed to decline in regions such as the African Sahel.

Climate change occurs when changes in Earth's climate system result in new weather patterns that last for at least a few decades, and maybe for millions of years.

3.4.3. Water Resources

Climate change is likely to substantially impact water resources. By the middle of this century, at present amounts of warming a 10–40% rise in mean river overflow and water accessibility has been anticipated in higher latitudes and in specific wet areas in the tropics, through reductions of alike amount are anticipated in other parts of the tropics, and in the dry areas of the subtropics, chiefly in summer. In many cases, water availability is expected to decrease or decreasing in areas which are now strained for water resources, for example, western North America, the African Sahel, Western Australia, and southern Africa. In such areas, drought is probable to rise in extent as well as magnitude, along with adverse effects (see above) the raising of livestock and on agriculture.

Earlier and Increased spring runoff is by now actually witnessed in extra-tropical areas with snow-fed or glacial streams and rivers, as like Western North America. Freshwater presently amassed by mountain snow and glaciers in both the extra-tropics and the tropics is anticipated to drop, decreasing freshwater accessibility by more than 15% of the Earth's inhabitants that depends on such freshwater sources.

It is also possible that warming temperature, over their influence on natural action in rivers and lakes, might have a negative influence on water quality, more lessening reach to safe water sources for farming and drinking. Risk-management actions are by now being engaged by several nations in reaction to anticipated alterations in water accessibility.

3.5. GLOBAL WARMING RESULTING IN VECTOR-BORNE DISEASES

Numerous communicable illnesses are stretched out by creatures as like rodents and mosquitoes, called as disease vectors, whose behavior and distributions are delicate to moisture as well as temperature. The risk of vector-borne disease could be increased by global warming in a number of ways:

Dengue fever is a mosquito-borne tropical disease caused by the dengue virus.

1. **Higher temperatures quicken the growth of specific disease-causing agents and their vectors:** For instance, the development period essential for a mosquito to be competent to spread **dengue fever** virus afterward it has been infested drops from 12 days at 30°C to 7 days at 32–35°C; it converts into a likely three times rise in the communication rate of ailment.
2. **Higher annual average temperatures can lengthen the season during which vectors are active:** In Canada, for example, environmentalists have discovered that with warmer autumns being the most likely cause, present-day mosquitoes delay nine days more than their ancestors used to 30 years before they initiate their winter inactivity.
3. **A warmer climate can expand the geographic range of tropical mosquito-borne diseases, as like dengue fever, malaria, and yellow fever, to higher latitudes and altitudes:** Particularly susceptible are less established nations which have scarcer resources to battle illness. The utmost widespread vector-borne disease, Malaria, takes toll of up to 2 million people per annum. The ailment is mostly restricted to areas with winter temperatures beyond 16°C, as the organism is not capable to breed below this temperature. Dengue diseases are usually restricted to the tropics between the latitudes of 30° north and 20° south, as prolonged cold weather or frosts destroy the adult Aedes aegypti mosqui-

toes, vector, and their over-wintering larvae as well as eggs.

As skeptics often emphasize, there are aspects additional to global warming, which may impact the communication of tropical illnesses. For instance, since 1973, there have been a revival of malaria mosquitoes all through the tropics, impacted more by the entrance of insecticide-resistant mosquitoes and by the malaria's parasites growing resistance to antibiotics rather than by global warming.

Global warming has far-reaching side-effects on several facets of social life. It intimidates lives, economies, and conventional styles of life. The evidences collected in this description displays a choice to the society: we can act now to stabilize the climate and mitigate future damages or we can completely ignore the incident of global warming, and in its place attempt to manage with its progressively distressing influences on our livings and the accepted earth we treasure.

REVIEW QUESTIONS

1. Define global warming.
2. What are the tradeoffs between possible benefits claimed for global warming?
3. Describe the destructive effects of sea level rise.
4. Explain observed changes in global, regional, and local climate.
5. What do you mean by the term "desertification"?
6. Illustrate some trends in stratospheric ozone depletion.
7. Define "Ecosystem"?
8. What are the regional climatic consequences of deforestation?
9. Define the impact of global warming on our society.
10. Why global warming leads to various vector-borne diseases?

REFERENCES

1. Caitlyn, R., Witkowski, J., Weijers, W. H., Brian, B., Stefan, S., & Jaap, S., (2018). Sinninghe Damst�. Molecular fossils from phytoplankton reveal secular Pco2 trend over the Phanerozoic. *Science Advances*,. doi: 10.1126/sciadv.aat4556.
2. *Global Warming – Environmental Consequences of Global Warming*. Retrieved from: https://www.britannica.com/science/global-warming/Environmental-consequences-of-global-warming (Accessed on 18 June 2019).
3. Mann, M., (2009). *Represent a Serious Threat to Our Welfare and Environment?* [eBook] (p. 38). Social Philosophy & Policy Foundation. Retrieved from http://www.meteo.psu.edu/holocene/public_html/shared/articles/MannSocialPhilos09.pdf (Accessed on 18 June 2019).
4. Nunez, C., (2019). *Causes and Effects of Climate Change*. Retrieved from: https://www.nationalgeographic.com/environment/global-warming/global-warming-overview/ (Accessed on 18 June 2019).
5. *Potential Impacts of Climate Change*, (1990). Retrieved from: http://www.ciesin.org/docs/001-011/001-011.html (Accessed on 18 June 2019).
6. Wang, J., & Chameides, B., (2005). *Global Warming's Increasingly Visible Impacts* [eBook] (p. 43). Environmental defense. Retrieved from: https://www.edf.org/sites/default/files/4891_GlobalWarmingImpacts.pdf (Accessed on 18 June 2019).

Web Resources

7. Intergovernmental Panel on Climate Change. https://www.ipcc.ch/
8. Resources about Climate Change. https://guides.lib.unc.edu/climatechange/home

Chapter 4

POPULATION GROWTH: BRINGING "LIFELESS" ENVIRONMENT ALONG

LEARNING OBJECTIVES:

In this chapter, you will learn about:

- The issues related to population growth.
- Impact of overpopulation on environment.
- Key causes of population growth.
- Ethical and conceptual issues of overpopulation.
- Specific arenas of population-environment interaction.

Key Terms:

- Environmental footmark
- Eutrophication
- Fossil fuel
- Global inequality
- Greenhouse gas
- Land use alteration
- Mortality rates
- Overpopulation
- Urbanization

4.1. INTRODUCTION

As the population of the world is increasing day by day, now it does not seem to pose any challenge for us for the last twenty years. However, this problem did not occur all of a sudden, rather it has been prevalent since the early histories of mankind and continues even today. All through times gone by, reputed researchers forecast the upcoming concerns if the sphere tracks a similar interactive model, they named it as "overpopulation."

By following the forecasts, researchers buckled down to invent various contraceptives and also used eugenics to control the world population. Even after that, the population could not be controlled and it kept on fighting with persistent illnesses. Another factor that aided in population increase further was immigration. It further imposed many dangers to the atmosphere as well.

Urbanization has led to the deterioration of natural climate and habitats of wildlife. It has contributed to high emissions of gases like carbon dioxide, which is responsible for global warming. Due to this, many species have become endangered, mankind itself is a threat. Out of many challenges, a few could be counted as: shortage of water, food scarcity, inadequate education and lack of job opportunities. Such problems then give rise to problems at a bigger scale like global inequality.

Figure 4.1: Over-crowded roads of Bangladesh.

Source: https://live.staticflickr.com/2145/2402764792_a960916575_b.jpg

It leads to irregular supply of natural resources, individual rights as well as financial means, which in turn gives rise to impoverishment. It blames the worldwide ethos as avaricious, in spite of the establishment of global agencies as well as organizations. The efforts of national and international agencies to execute the guidelines provided by global agencies to improvise the state of mankind in general are the best solutions to overpopulation (Figure 4.1).

Individuals take action what suits them best without paying any heed to the acute shortage and interest within the global network. Issues, which add to abrupt increase in population as well as extinction of **human values** and ethics, are reinforced by inequality.

Human values refer to those values which are at the core of being human.

Planet Earth is a bounded system and the earth is provided with limited natural resources. For every new child born, environment suffers in terms of scarcity of resources. Majority of people incline to concentrate to achieve their own short-term self-goals for contentment and pleasure thereby, completely ignoring the protection and conservation of ecosystem for long-term objectives. It is vital for human beings to comprehend that we are going to get any outside help which can solve the problem of shortage of natural resources.

We the residents of earth need to understand the sensitivity of the issue which we will experience very soon in the near future. Ecological campaigners recognize the concern, whereas people other than those think that they would be able to find a solution to this problem as all other problems related to environment have been resolved in the times gone by. It is very difficult to make people understand that now nature has taken an irretrievable direction and all the harmful events occurring now, cannot be reverted to the way as they were before.

In all the species, events of overpopulation occur; it is not confined to mankind only. However, we are intelligent enough to save ourselves from the natural calamities that cause mass deaths like epidemics, drought, tsunami, etc. Commonly found reasons of overpopulation are listed below (Figure 4.2):

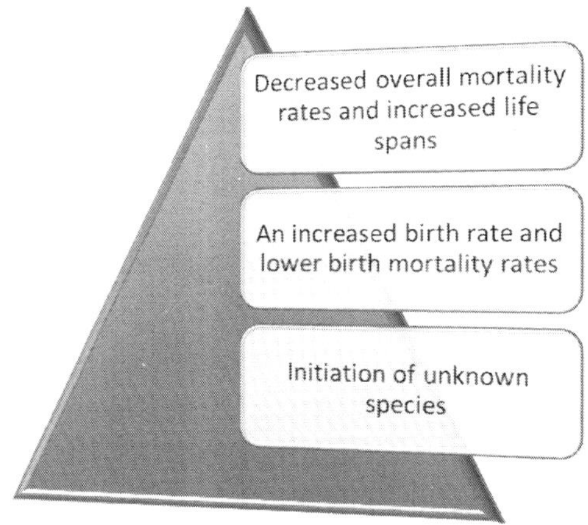

Figure 4.2: Major reasons behind overpopulation.

- Decreased overall mortality rates and increased life spans.
- An increased birth rate and lower birth mortality rates.
- Initiation of unknown species – does not usually relate to human being.

Increased birth rate, lower overall mortality rate, and improved life expectancy are some of the important results of innovations in scientific research and dramatic improvement in medical sciences.

4.1.1. Conceptual and Ethical Issues

We have two contrasting schools of thought on the environment and overpopulation growth. Intellects on one side are trying to find out the way growth of population affects the ecological modification; on this side debates are on-going to devise the best measure and idealize the complicated relationship between development, population, as well as the atmosphere.

Second, school of thought includes dissimilarities in principles on the principal ethical concerns that are the basis for the policy objectives: what kind of world should we be aiming for and what is to be valued. In terms of economics, how social welfare function should be specified in an appropriate manner? Such important, essentially moral problems are usually not answered, which lead to a lot of confusion among the two schools of thought.

To some extent, moral concerns regarding the crucial guideline's goals are entwined with the way one conceptualizes the relationship among mankind and the ecosystem. It can be best explained by the following example: we can always find a stark contrast amid the two schools of thought which are of utmost importance in exploration of the environment and population: physical scientists – biologists, climatologists, etc. and economists (along with many other researchers in the social sciences).

Such differences include both fundamental ethical issues as well as scientific issues, with economists inclined to highlight and authenticate their equations and models on the potential of market price mechanisms to efficiently assign the usage of resources with time and when resources become scarce technological changes need to be promoted.

Human capital is the stock of habits, knowledge, social and personality attributes (including creativity) embodied in the ability to perform labor so as to produce economic value.

Economics feel a risk in overgeneralizing their opinions, because the atmosphere is important primarily as a basis of natural resources for manufacture and for its absorptive capability in relation to treating discarded remains from both economic manufacture and consumption.

These natural resources are measured on par with physical and human capital. When they are united with technology, they yield the substantial wellbeing for the mankind. Though there are many people who value to preserve the environment for many reasons, such individuals get contentment from the act of conservation or aesthetically from "environmental resources" as like parks.

So, for this reason, the people who identify the advantages of conservation should be made to bear the expenses of such undertakings.

Looking at it from the viewpoint of maximum environmentalists and natural scientists, indeed all-human activities and the economy are as a substitute observed as happening in the broader ecosystem. There are two major implications of this perception:

- Firstly, the environment is more than simply a set of natural resources. It is viewed as the critical prerequisite for long-term economic growth and improvements in human welfare, and not as merely one of numerous contributions to the manufacture procedure.

- Secondly, labors to conserve the environs is of worth further than the bounds of their influence on social wellbeing, as the non-human species and the biosphere do have an inherent worth of their own.

Let's consider an instance, contemplate on the incident of the other apes, they have maximum of man's genetic structure, display at least a rudimentary ability to reason as well as clearly feel physical pain and emotions.

Is it appropriate to encourage their preservation in zoos or their use for testing drugs and doing experiments by medical laboratories, today animal rights organizations want to prohibit those testing. However, reputed organizations in England and the US indicate that advances in sciences were not possible without the use of animals for experimentation in labs. Should additional steps be taken to assure their conservation as they characterize a wealth trove of behavioral and genetic data that can be evidence of high worth to mankind in the forthcoming times? Does the fact yet poorly understood, there may occur interrelations among all species living together in the ecosystem, possibly creating their future and our own someway related additionally upsurge the worth of their existence to us?

4.2. DIMINISHING FORESTRY IN THE EMERGING BIOSPHERE

Basic to the development of human civilization are the forests of our planet. Forest ecosystems seen as a kind of resource unique in itself, are considered in forestry as an amalgamation of multidimensional usefulness and effortless renewal ability by management of populations or planting of new individuals. For some species of slow growth, however, they might be species in risk of local extinction, and that is why nowadays many forest tree species are present in the red books of plants.

Figure 4.3: A decline in forestry has been witnessed from the past few years.

Source: https://www.cia.gov/library/publications/the-world-factbook/attachments/images/large/GB_001_large.jpg?1528323814

Along with the existence of humanity, the forest species have been fundamental for human settlements in fuel, housing, furnishings, paper, and packaging, wood products and constitute a major source for commodities in the global economy (Figure 4.3).Other useful non-wood products from plants are nuts, berries, fruits, and therapeutic herbs. For instance, as high as 90% of the highly prescribed pharmaceutical drugs in the USA are derived from animals and forest plants. Wood and its merchandises stand on the third position in worth amongst the sphere's merchandises, leaving behind only oil and natural gas, in spite of the fact that wood has been replaced for many traditional uses by coal, oil, steel, and plastics.

Accompanying their straight economic advantages are the unique environmental tasks of the biosphere's forests. Biological diversity finds its highest expression in tropical forests. At least half of the world's known species are sheltered in the tropical forests alone.

Fossil fuel is a fuel formed by natural processes, such as anaerobic decomposition of buried dead organisms, containing energy originating in ancient photosynthesis

Forests provide erosion control against landslides and flooding, protect, and enrich soils and watersheds, regulating the quantity of water and enhancing water quality. Forests also help to mitigate potential climate change; the world's forests help to offset anthropogenic emissions from burning **fossil fuel** as a major sink for carbon dioxide.

Other important aspects are the esthetic, recreational, as well as religious values of forests. The rates of destruction have been highest in recent decades, and further forests have been cut from 1850 as compared to all of preceding past. Though forest cover in the industrialized biosphere is intensifying these days, in the developing nations, this has been more than counterbalanced by devastation of forests. Continents like Africa and Asia, areas with the highest population growth, have vanished approx. assessed 65 to 70% of their original forest cover.

It has been hard to offer decisive evidence of the connection among forest destruction and population growth because of the difficulties intricate in human use of forests. Nonetheless, a fresh study conducted on more than hundreds of nations by a global survey agency determined that changes in population are the reason for nearly half of the deforestation over the span of mankind past.

1.7 billion people in forty countries already were, it has been assessed that by the year of 1995, because of vulnerable to shortages of fuelwood, inadequate forest cover, and to rising influences of erosion and flooding. Since then, merely in 30 years, by the year 2025, this figure is anticipated to increase to 4.6 billion people in 53 nations.

During the 1980s, it was revealed by World Bank that while having the lowest per capita income and the highest population growth rate, Central Africa underwent the world's second highest rate of deforestation. Haiti is also the poorest and most densely populated as well as the most severely deforested nation in Central America. The U.N. Food and Agriculture Organization, in South Asia, cautioned that maintenance hard work "will be invalidated and in places upturned, without escorted by a decrease in the rate of growth of population."

4.3. TWO SPECIFIC AREAS OF POPULATION-ENVIRONMENT INTERACTION

Global land-use patterns and climate change are the two particular arenas of analysis that need assistance to validate the problems of considering the multifaceted effect of population dynamics on the environments.

Conversely, the interrelations amid environmental context and demographics are illuminated by the examples which validate the rising body of scientific proof.

4.3.1. World-Wide Climate Change

On record, recent years have been amongst the warmest. Data proposes that hotness have been affected by rising concentrations of greenhouse gases (GHGs), as like carbon dioxide, which has a property to captivate solar energy that warms up the planet's environment. How far should the climate change be credited straight to demographic aspects? There is an enormous load of results from investigations and proofs which claim that most of alterations in atmospheric gas are induced by humans.

The demographic impact seems chiefly in three procedures: additions to carbon dioxide releases occurring from use of fossil fuels linked to energy consumption and industrial production; further consumption related procedures, as like livestock production and rice paddy cultivation, are accountable for GHG emissions to the air, mainly methane; and land-use changes, such as deforestation, also affect the exchange of carbon dioxide between the Earth and the atmosphere.

The investigation has exposed that population growth and size are significant elements in the release of GHGs. Another research proposes that population growth and size will contribute to approximately 35% of the worldwide upsurge in **Carbon dioxide** releases during the period of 1985 to 2100.

Carbon dioxide is a chemical compound composed of one carbon and two oxygen atoms.

Additionally, during that period, 48% of the upsurge would be from emerging countries. Though, as forecasted, population growth is going to slow down in the coming years, its involvement to releases of gases is also expected to drop.

Such drop-in GHGs would be particularly contributed by the emerging countries. Though releases of carbon dioxide from the emerging countries are assessed to account for 42% during the time span of 1985 to 2020, they are estimated to account for merely 3% during the time span of 2025 to 2100.

4.3.2. Land Use

Accomplishing the requirements of resource of an increasing population eventually necessitates certain kind of transformation in land-use, for instance, to intensify production on already cultivated land, expand food production through forest clearing, or to improve the arrangement required to upkeep bigger population. Certainly, the rapid pace of contemporary population growth is because of the man's skill to control the land use.

Two noticeable practices of man encouraged land use alteration are agriculture and deforestation. From the last 300 years, increasing from 2.65 million square kilometers to 15 million square kilometers, the expanse of globe's land used for agriculture has increased to approximately 450%. The forest land in the world is reducing day by day at the same time. Because it often represents a consequence of agricultural expansion, deforestation is strictly related to agrarian land use alteration.

Although changes in forest cover vary greatly across regions, total decrease in forest land of 180 million acres happened through the period of 15 years, i.e., 1980 to 1995. Deforestation and changing land use create numerous environmental influences in particular. Agriculture gives rise to the problem of soil erosion, at the same time uncontrolled use of fertilizers and chemicals inputs degrades the soil too.

Deforestation too leads to increase in soil erosion, along with lessening the ability of soils to hold water, reducing rainfall due to localized climate changes, as well as growing the severity and frequency of floods. In general Land use alteration ends in loss of habitat and fragmentation (which is the main reason for extinction of species). Researchers claim that if the present rate of deforestation endures, nearly 25% of all the species on the planet may get extinct in the coming 50 years.

4.4. POPULATION CONSUMPTION AND ENVIRONMENT DEGRADATION

It has been clearly proven that environmental degradation and poverty connected closely, it is of even greater concern predominantly in developed nations, the unmanageable patterns of production and consumption.

Usually, people in western nations never stop to consider the level of consumption in developing countries. Particularly in industrialized nations, for many, the consumption of resources and goods has unprejudiced turned out to be a share of our culture and lives, encouraged by both the advertisers as well as by the governments, desirous to constantly develop their economy. Traditionally, all through the globe it is taken as a normal behavior to buy, shop, and utilize, to persistently attempt to possess a faster car or a finer home, these all are often treated as symbols of prosperity.

Consumer Culture focuses on the spending of the customer's money on material goods to attain a lifestyle in a capitalist economy.

Though it is perfectly okay to demonstrate the **consumer culture** and to appreciate the worldly belongings, but excess to everything is bad and so is the case here. It is harming both our emotional wellbeing as well as the planet.

The ecological influence of all this utilization is vast. The bulk manufacture of merchandises, numerous of them totally needless for a contented and relaxed life, is consuming great quantities of energy, generating huge amounts of waste and creating excess pollution.

All the system is so interrelated and complicated that the ecological influences of great levels of utilization is not limited to a small area or district and not even a country, for that matter. For instance, the usage of fossil fuels for generation of energy (heat and cool our bigger houses, to drive our bigger cars) has an influence on worldwide Carbon dioxide levels and thus ensuing ecological impacts.

In the same way, wealthier nations are also competent to depend on waste-intensive imports and/or resource being created in poorer nations. So, they are saved from dealing with the immediate impacts of the factories or pollution that went into creating them, and they are able consume and enjoy the products.

From a global perspective, all human being is not equally accountable for ecological damage. Resource use and consumption patterns are extremely huge in several regions of the earth, whereas in the rest of the world, mostly in highly populated countries they are insignificantly low such that even the elementary necessities of all the inhabitants could not be fulfilled (Figure 4.4).

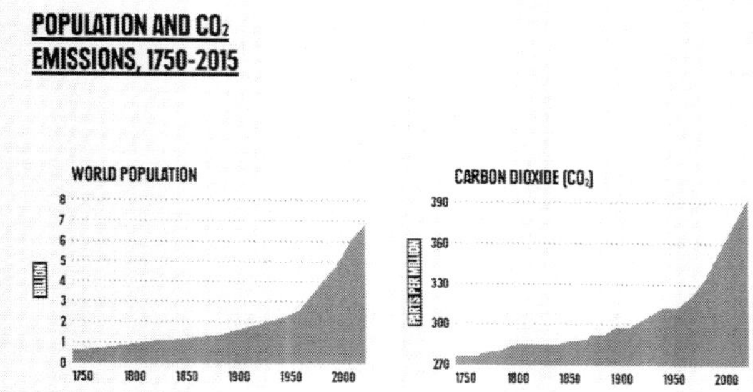

Figure 4.4: Population growth and carbon emissions.

Source: United Nations 2017

According to current studies, on average, the nations which had the slowest increases in carbon emissions are the ones with fastest growing population during the period 1750 and 2015.

Folks dwelling in advanced nations have, all together, a considerable superior environmental footprint as compared to those dwelling in the emerging nations. The 10 countries with the highest carbon footprints are United States > Canada > South Korea >Russia> Japan > Germany >China> United Kingdom > Italy > France >Brazil>India.

4.5. GLOBAL WARMING AND EFFECTS OF OVERPOPULATION

Climate change and human population growth have developed together by way of the usage of fossil fuels blowup to upkeep industrialized societies. "More people means more demand for gas, oil, coal, and other fuels drilled or mined from below the surface of the earth which when flamed, emit sufficient carbon dioxide (CO_2) into the environment to entrap warm air inside just like a greenhouse," records Scientific American. Developed countries consume most of the fossil fuel. It is no less than a modest opinion that maximum emerging countries aim to comparable industrialized economies as they encounter economic growth, which furthermore worsens Carbon dioxide releases into the air.

Another important component of GHG emission is deforestation. Worldwide, forests are capable of storing more than double the quantity of Carbon dioxide that is present in the air. As forests are burned and cleared, that carbon dioxide is emitted into the air, contributing to approximately 25% of the entire GHG emission.

The association amid environmental impacts and overpopulation are frequently complex and interrelated. Several of the important sustainability challenges related to overpopulation are mentioned below, focusing on carbon sequestration and release.

To understand these challenges in a simplified way, they have been listed discretely; however, the fact is that they are interrelated and complex in nature and hence managing them is all the more challenging.

4.5.1. Farming Impacts

A rising agrarian base to nourish the increasing population of the world has many other problems. As the total population goes on increasing, requirement for food increases further. These demands could be fulfilled through deforestation to create new farmlands or through more intensive farming, which then again poses adverse consequences. Worldwide agriculture has contributed to approximately 80% of deforestation.

Through intensive farming, the production can be increased in order to fulfil the requirements of gradually increasing population.

However, this method is characterized by dependence on pesticides, mechanization, and synthetic fertilizers.

These perform then give rise to the problems of depletion or soil erosion. The land abandoned and used in the previous 50 years worldwide is nearly equal to the amount of land in use currently. Additionally, the agrarian overflow of surplus fertilizers is the chief reason of the problems like **eutrophication**, which reduces oxygen from water and consequences in substantial side effects for aquatic life. Pesticides and herbicides are major solutions for pests and reduction of labor with extreme consequences as non-point pollution of waters. Damaging of soil increases CO_2 to the atmosphere and loss of carbon with erosion.

Eutrophication is when a body of water becomes overly enriched with minerals and nutrients which induce excessive growth of algae.

Increasing nowadays are large scale developments of organic agriculture, sustainable use of local materials, and recycling of organic wastes, especially in developed countries such as the US and France. Management of soil and composting of manure have a positive impact on carbon cycle as it reduces CO_2 release to the atmosphere

4.5.2. Deforestation

The problem of GHG emissions is exasperated due to deforestation. Sequestration of carbon in tropical forests account for more than 90% of the whole carbon in trees worldwide. Deforestation reduces the capability to capture carbon dioxide and also the generation of oxygen for the planet. In South America only, Tropical rainforests are accountable for generating 20% of the total earth's oxygen.

As mentioned above, agriculture is accountable for around 80% of worldwide deforestation. Among the rest, another 14% is credited to logging, 5% to collection of firewood, and the balance resulting from other causes. Human population upsurge is connected to all of such deforestation burdens. More population implies people will need more wood products, more food as well as more firewood.

4.5.3. Eutrophication

One of the main causes of eutrophication in waters derives from non-point pollution increasing nutrients such as N, P, and Si in the water bodies generated as Agrarian overflow: eutrophication leads to lack of oxygen in the water profile causing dead zones where species are displaced from their original habitats both in freshwaters and also in marine environments.

Another main source of eutrophication is sewage and industry disposal--both associated to population growth.

On a global level, eutrophication has caused more than 400 marine 'dead zones', jointly wrapping the region six folds the size of Switzerland. A remarkable example can be seen increasing every year the surface in the Dead Zone of the Gulf of Mexico with high mortalities of fish and other aquatic marine organisms.

Eutrophication results in the death of aquatic animals caused due to the high growth of plant life that lives on oxygen. Accumulation of carbon in sediments is a positive result of eutrophication in terms of climate change; however, it also produces local affectation of biodiversity and increase in acidity (Figure 4.5).

Figure 4.5: Eutrophication.

Source: https://s0.geograph.org.uk/geophotos/03/78/40/3784065_47969c09.jpg

The usage of fossil fuel for electricity production to power industrial units also generates nitrogen oxides (NOx), sulfur dioxide (SO_2), and carbon dioxide (CO_2), which can finally be captivated by water bodies to upsurge their nutrient capacity.

REVIEW QUESTIONS

1. Describe the impact of growing population on pollution.
2. Explain urbanization.
3. What are the major causes of overpopulation?
4. What are the ethical issues related to the population growth?
5. Why forests are gradually declining in the developing world?
6. Define two significant areas of population and environment interaction.
7. How population consumption is related to environmental degradation?
8. Briefly describe environmental effects of overpopulation.
9. Define the term "Eutrophication."
10. How the effect of global warming can be reduced?

REFERENCES

1. Baus, D., (2017). *Overpopulation and the Impact on the Environment* [eBook] (p. 60). CUNY Academic works. Retrieved from: https://academicworks.cuny.edu/cgi/viewcontent.cgi?article=2929&context=gc_etds (Accessed on 18 June 2019).
2. Benedick, R., (2000). *Human Population and Environmental Stresses in the Twenty-First Century* [eBook] (p. 14). Retrieved from: https://www.wilsoncenter.org/sites/default/files/Report6-1.pdf (Accessed on 18 June 2019).
3. Hunter, L., (2000). *The Environmental Implications of Population Dynamics* [eBook] (p. 123). RAND. Retrieved from: https://www.rand.org/content/dam/rand/pubs/monograph_reports/2000/MR1191.pdf (Accessed on 18 June 2019).
4. Leblanc, R., (2018). *The Environmental Impacts of Overpopulation*. Retrieved from: https://www.thebalancesmb.com/how-overpopulation-impacts-the-environment-4172964 (Accessed on 18 June 2019).
5. LeGrand, T. *World Population Growth and the Environment* [eBook] (p. 6). Encyclopedia of life support systems. Retrieved from: http://www.eolss.net/Sample-chapters/C11/E1-10-04-01.pdf (Accessed on 18 June 2019).
6. Nichols, M., (2018). *How Does Overpopulation Effect the Environment? | Schooled by Science*. Retrieved from: https://schooledbyscience.com/overpopulation-effect-environment/ (Accessed on 18 June 2019).
7. *Population and Environment: A Global Challenge*. Retrieved from: https://www.science.org.au/curious/earth-environment/population-environment (Accessed on 18 June 2019).

Chapter 5

HUMAN HEALTH AND ENVIRONMENTAL ISSUES

LEARNING OBJECTIVES:

In this chapter, you will learn about:

- The relation between human health and environmental issues.
- Environment-related health problems.
- The concept of urban sanitation in context to environmental pollution.
- Effects of environmental humiliation on human health.

Key Terms:

- Chronic obstructive pulmonary disease
- Dehydration
- Diarrheal disease
- Endothelial dysfunction
- Endotoxins
- Ergonomic issues
- Respiratory disease
- Vector-borne diseases

5.1. INTRODUCTION

Environmental pollution is the most severe problem the humanity, and other life forms are facing today on our planet. Due to uncontrolled urbanization and industrialization, environmental degradation has been occurring very rapidly and causing many problems like land insecurity, worsening water quality, excessive air pollution, noise, increase in energy consumption and the problems of waste disposal.

Despite the major efforts that have been made over recent years to clean up the environment, environmental pollution is one the major problems that affect biodiversity, ecosystems, and human health worldwide by contaminating soil and water and poses continuing risks to health. The impact of pollution is more severe in developing countries, leading to ill health, death, and disabilities of millions of people annually.

In the developing world, the traditional sources of pollution such as industrial emissions, poor sanitation, inadequate waste management, contaminated water supplies and exposures to indoor air pollution from biomass fuels affect large numbers of people. Developed countries have the resources and technologies to combat pollution, and efforts have been made to reduce pollution. There it majorly persists amongst the poorer sections of the society. Over the past decades, various sources of pollution were identified that altered the composition of water, air, and soil of the environment. The substances that cause pollution are known as pollutants. Because of the vehicular exhausts and use of modern chemicals in the home, in food, for water treatment and for pest control, a wide range of modern pollutants have emerged. The pollutants are present in lower concentrations, so the ill-effects in health are not immediate or obvious, but they imply large relative risks in the future. Detecting small effects against a background of variability in exposure and human susceptibility, and measurement error, poses severe scientific challenges. The greenhouse gas (GHGs) emissions, acid depositions, water pollution, waste management, and global environmental pollution are together considered as an international public health issue. These factors influence the health of a population, including diet, sanitation, socio-economic status, literacy, and lifestyle (Figure 5.1).

Figure 5.1: Acid deposition and climate change.

Source: https://www.nps.gov/articles/images/2852A505-1DD8-B71C-0764CD-6D2A0E6B7BHiResProxy.jpg?maxwidth=650&autorotate=false

Thus, for enhancement of our physical, mental, and social well-being, they should be investigated from multiple perspectives, including social, economic, legislation, and environmental engineering systems, as well as lifestyle habits. This may help in strengthening environmental systems to resist contamination and therefore promoting health. Environmental pollutants have various adverse health effects from early life, which may vary from allergies to cancer. Even at low exposure levels, urban air pollutants can cause asthma, allergies, respiratory diseases, and **cardiovascular diseases** if the exposure is continuous or long-term.

Cardiovascular disease generally refers to conditions that involve narrowed or blocked blood vessels that can lead to a heart attack, chest pain (angina) or stroke.

Other harmful effects are perinatal disorders, infant mortality, malignancies, increase in stress oxidative, endothelial dysfunction, mental disorders, etc. Heavy metals have been shown to cause neurological disorders and various cancers. Usually, short-term effects which pose immediate implications are highlighted but the need of the hour is to detect the pollutants from early life and highlight their possible implications on chronic non-communicable diseases of adulthood. Numerous studies have suggested that environmental particulate exposure has been linked to increased risk of morbidity and mortality from many diseases, organ disturbances, cancers, and other chronic diseases. In addition to physical diseases, environmental contamination can also cause psychological problems. Noise can have a direct impact on human health decreasing the quality of life and potentially contributing to depression. Therefore, it is time to take action and control the pollution. Otherwise, the waste products from consumption, heating, agriculture, mining, manufacturing, transportation, and other human activities will continue contributing to widespread release of chemicals in the environment and thus degrade the environment.

5.2. LINKS BETWEEN ENVIRONMENTAL POLLUTION AND HEALTH

Environmental pollution can be simply be defined as the presence of an agent in the environment which contaminates the biological and physical components of the environment to such a big extent that it is potentially damaging to the normal environment processes and thus human health. As such, pollutants take many forms. They can be naturally occurring substances or energies, but when in excess of natural levels, they are considered contaminants. They include not only chemicals, but also organisms and biological materials, as well as energy in its various forms (e.g., noise, radiation, heat). Any use of natural resources at a rate higher than nature's capacity to restore itself can result in pollution of air, water, and land. The number of potential pollutants is, therefore, essentially countless.

There are some 30,000 chemicals in common use today, any one of which may be released into the environment during processing or use.

Out of such a large number of known pollutants, even less than 1% has been subjected to a detailed assessment in terms of their toxicity and health risks. In reality, it is impossible to express the biological pollutants in terms of quantity. They include not only living and viable organisms, such as bacteria, but also the vast array of **endotoxins** that can be released from the protoplasm of organisms after death. Thus, they pose a great potential environmental risk to health and it has become of utmost importance to understand the nature and mechanism of these risks.

5.2.1. The Source–Effect Chain

Endotoxins are part of the outer membrane of the cell wall of Gram-negative bacteria.

Long-term exposure to pollution can result in significant health problems. Pollution is linked to health in a complex and unpredictable way. Pollutants cause adverse effects on human health and the environment. To have an effect on health, susceptible individuals should receive sufficient doses of the pollutant, or its decomposition products to trigger detectable symptoms. Actual risk of adverse effects depends on individual's current health status, the pollutant type and concentration, and the length of the exposure to the polluted air.

When individuals have been exposed to pollutants either repeatedly or over a long period of time, the damage occurs, and the symptoms start appearing. This happens when the susceptible individuals and pollutants shared the same environments at the same time.

For this kind of exposure, it is required to expose the humans to the environment in which the pollutants have been released and dispersed. Once in the environment, pollutants may be dispersed via air, water, soil, living organisms, and food.

The path of their dispersion, the rate, and pattern of dispersion depends on both the source and emission and the pollutant concerned and to a large extent, on environmental conditions. This complexity makes it is often very difficult to measure pollutant patterns and trends, and thus to predict levels of human exposure.

Therefore, the health hazard caused by the environmental pollutants is totally unavoidable. They depend on the coincidence of both the emission and dispersion processes that determine where and when the pollutant occurs in the environment, and the human behaviors that determine where and when they occupy those same locations.

5.2.2. Atmospheric Emissions

Rapid urbanization means that we are now exposed to unhealthy concentrations and a more diverse variety of ambient air pollutants. Air pollution measurements give important, quantitative information about ambient concentrations and deposition and can describe air quality at specific locations and times. Air pollution models are the only method that quantifies the complex relationship between emissions and concentrations/depositions, including the consequences of past and future scenarios and the determination of the effectiveness of abatement strategies.

This makes air pollution models necessary in regulatory, research, and forensic applications as they help in the existence of better-established policy and regulations. Almost all economic and societal activities result in emissions of air pollutants. There are many different types of sources of atmospheric emissions, for example, industry, energy supply (power plants, refineries, incinerators, factories, fossil fuel extraction and production sites, etc.) transport, animals, and humans, agricultural activities, natural, and managed vegetations, soil, etc.

One of the most important emission processes for many pollutants is the combustion from industrial sources as well as from low-level sources such as motorized vehicles and domestic chimneys, and indoor sources such as heating and cooking in the home or workplace. The emission from the industrial combustion or waste incineration happens at very high temperature and through tall stacks.

Due to this, they are dispersed widely within the atmosphere as compared to the emissions from low- level sources such as road vehicles and low-temperature combustion sources such as domestic heating. These less widely dispersed pollutants result in high pollution concentration, creates steep pollution gradients in the environment and contributes to creation of local pollution hot spots.

For example, in urban environments, high pollution concentration is found on busy roads as traffic-related pollutants such as nitrogen dioxide and carbon monoxide typically show order-of-magnitude variations in concentration over length-scales of tens to a few hundred meters. Some other important emission processes contributing to local variations in environmental pollution includes evaporation and leakages.

Given the link between environmental problems and human health, more of us are realizing that the initial investments to curb environmental issues and the exorbitant up-front costs for environmental clean-up ultimately pays off well in the long run as the health care costs goes down our loved ones lives longer, healthier lives.

There are free websites like Environmental Health News offered by the nonprofit Environmental Health Sciences which publish daily articles related to man-made environmental problems and the way in which they are linked to the humans and how they directly/indirectly pose impact on public health. Even the local TV station or newspaper likely carries an occasional story about the health effects of environmental pollution. The commitment of younger generations to a cleaner future and the new found public awareness regarding the environmental issues is promising and suggest that we are moving in right direction.

5.3. MAJOR ENVIRONMENTALLY RELATED HEALTH PROBLEMS

5.3.1. Diarrheal Disease

Diarrheal diseases are major public health problems, especially in children in developing countries. Globally, diarrhea remains the second most common cause of death among children under five years of age. It is clinically defined as three or more loose stools passed over a twenty-four-hour period. Diarrhea can last several days and can leave the body without the water and salts that are necessary for survival.

Diarrhea is usually a symptom of an infection in the intestinal tract, which can be caused by a variety of bacterial, viral, and parasitic organisms. Infection is spread through contaminated food or drinking-water, or from person-to-person as a result of poor hygiene. It is also a common symptom of a range of other diseases that are not related to stomach infections, such as malaria, measles, respiratory infections, and AIDS. Many times, it is characterized by the presence of blood in stools, i.e., dysentery, but not necessarily excessively loose or frequent. Severe **dehydration** and fluid loss can lead to diarrhea deaths, particularly among children, and to impaired growth and nutritional status among the survivors.

Dehydration is a condition that can occur when the loss of body fluids, mostly water, exceeds the amount that is taken in.

Most of the cases of diarrhea worldwide are the result of fecal-oral contamination. Water supply, sanitation, and hygiene are the top proven preventive interventions for reducing diarrheal disease transmission.

5.3.2. Respiratory Disease

Respiratory infections are the type of disease that affects the lungs and other parts of the respiratory system and includes the lungs, throat, mouth, and middle ear tract. Respiratory diseases include asthma, chronic obstructive pulmonary disease (COPD), pulmonary fibrosis, pneumonia, and lung cancer. Acute respiratory infection (ARI) is a serious infection that prevents normal breathing function. It prevents the body from getting oxygen and can result in death.

Also, ARIs are infectious, which means they can spread from one person to another. The disease is quite widespread and imposes the largest burden of disease globally. It is particularly dangerous for children, older adults, and people with immune system disorders and is the leading cause of death in the developing world. The early symptoms of ARI usually appear in the nose and upper lungs and leads to minor upper respiratory infections, such as colds and sore throats, ear infection and tonsillitis, as well as more serious lower respiratory infections, such as pneumonia, croup, and bronchitis.

Most ARI episodes are mild and do not require specific treatment, but in some unfortunate cases, it becomes severe and progress into life-threatening pneumonia, which is the biggest cause of child mortality. The immune systems of children and the elderly are more prone to be affected by viruses. People with heart diseases or other lung problems are more likely to contract an ARI. Anyone whose immune system might be weakened by another disease is also at risk.

The fatality rate is higher among those with poor nutritional or socio-economic status. The victims of tuberculosis majorly are the adults who are the wage earners of the family, which results in high social and economic costs of tuberculosis. Thus, Tuberculosis can drive entire countries, not just the people within them, into ever deeper poverty. It is a fact that countries with the highest rates of TB are also some of the poorest.

The disease is associated with poverty because people are more likely to live and work in poorly ventilated and overcrowded conditions, which provide ideal conditions for TB bacteria to spread; and suffer from malnutrition and disease which reduces resistance to TB; and have limited access to healthcare. The affected families are usually those with the least means to cope with the death or debilitation of an adult wage earner. Loss of productivity associated high costs of treatment and the attached stigma worsen the condition of already poor people.

5.3.3. Malaria and Other Vector-Borne Disease

Malaria is a mosquito-borne disease spread to people through the bite of an infected Anopheles mosquito. Other vector-borne diseases include Dengue, Lymphatic filariasis, Chagas, Trypanosomiasis, and Leishmaniasis.

The distribution of all vector-borne diseases is based on the environmental factors, such as water, temperature, humidity, vegetation, and the structure of the built-up environment. Poorly designed irrigation and water systems, inadequate housing, poor waste disposal, and water storage, deforestation, and loss of biodiversity, are the contributing factors to the most common vector-borne diseases including malaria, and dengue. The growth of urban slums, lacking reliable piped water or adequate solid waste management, can render large populations in towns and cities at risk of viral diseases spread by mosquitoes.

Vector-borne diseases are human illnesses caused by parasites, viruses and bacteria that are transmitted by mosquitoes, sandflies, triatomine bugs, blackflies, ticks, tsetse flies, mites, snails and lice.

Generally, **vector-borne diseases** tend to decrease with urbanization as the certain vectors, such as Culex mosquitoes, carriers of filariasis, have important breeding sites in blocked drains and waterlogged latrines, the situation associated with unplanned and poor urbanization.

The most important vector-borne disease is malaria, being the third largest 'environmental contributor' to the global burden of disease. Malaria is endemic in 104 countries and affects half the world's population. More than 660,000 deaths are recorded each year, and approximately between 300 and 500 million people become acutely ill.

Almost 90% of these deaths occur in Sub-Saharan Africa, especially in children under five. Pregnant women and their unborn offspring are also particularly at risk from malaria, which is a major cause of infant death, low birth weight, and maternal anemia. The risk of severe malaria is limited to people who are not immune.

With the increase in severe malaria in adults, people are now taking personal precautions and preventative measures such as insecticide-treated mosquito nets, indoor residual spraying, and education on malaria. They are increasingly live in relatively protected urban environments, and hence delay the development of immunity. Those adults who live in areas where malaria is seasonal, and who therefore do not develop immunity, are at utmost risk. Besides, travelers and migrants originating from non-endemic areas are also prone to get subjected to the disease.

5.4. URBAN SANITATION AND ENVIRONMENTAL POLLUTION

Globally there is a burden of disease associated with environmental pollution hazards which include infectious diseases related to drinking water, sanitation, and food hygiene; respiratory diseases related to severe air pollution; and vector-borne diseases such as malaria. Liquid waste management, solid waste management, and outdoor air pollution have emerged as the major risks and challenges related to environmental pollution.

Wastewater generation has increased due to increase in population, urbanization, and industrialization. The sewage system breakdown in the cities of Blantyre and Lilongwe and to a lesser extent, Mzuzu, and Zomba are the major causes of water resources pollution. Sewage system break down, sewer lines blockages occur due to poor maintenance (vandalism of pipes), improper design of some sections (vandalism of pillars that support pipes, high load for pipes, old pipes), and also lack of public awareness on use of the sewerage systems and carelessness on what to throw in a water closet latrine.

The other causes include discharge of waste (industrial and general) into storm drains, roadsides, streams, and rivers; sand mining, farming along the catchment area and defecation along the riverbanks. **Wastewater** contains chemical pollutants such as heavy metals, pathogens, and helminths, such as roundworms, hookworms, and guinea worm that threaten the health of humans as well as the environment.

Wastewater is "used water from any combination of domestic, industrial, commercial or agricultural activities, surface runoff or stormwater, and any sewer inflow or sewer infiltration".

The worst-case situation occurs when untreated wastewater is used to irrigate vegetables or salad crops that are eaten raw. This affects the urban dwellers who buy vegetables and other products grown and/or washed by this polluted water while the poor people are the most affected ones as they use this contaminated water for domestic household purposes.

So, the lack of adequate wastewater treatment causes severe water pollution and outbreaks of waterborne diseases. Polluted water needs to be dully treated in order to minimize its negative effects on people, animals, birds, and aquatic life. Despite such existing problems, the existing regulatory framework, institutional arrangements, and policy guidelines underscore the importance of wastewater treatment. With an increase in the global population and the rising demand for food and other essentials, there has been a rise in the amount of waste being generated daily by each household. Solid waste disposals are seriously spoiling the environmental conditions in developing countries. Negative environmental impacts from improper solid waste management are being observed throughout the world. Improper solid waste management leads to environmental pollution and contributes to pollution of water bodies as well.

The emerging risks from solid waste are mainly due to nature of the waste. *Exposure to hazardous waste* can affect human health, children being more vulnerable to these pollutants. Hazardous wastes may be inflammable, corrosive, explosive, toxic, mutagenic, carcinogenic, and eco-toxic. In other words, they have properties that make them dangerous or potentially harmful to human health or the environment. The release of chemical waste into the environment leads to chemical poisoning and thus diminishes public health safety when improperly managed.

They pollute water, are highly corrosive, and have a substantial or potential threat to public health. Hazardous waste cannot be disposed of by common means like other by-products of our everyday lives. If discharged in waterways, it pollutes the drinking water and is a big threat to aquatic life. If discharged on landforms, it makes the arable land unsuitable for farming and destroys vegetation.

The electronic waste contains chemicals

that are reactive in nature; thus, they are much more dangerous. It includes the waste from electronic products like computers, computer batteries, phones, and their batteries. Besides these, plastics, diapers, and other sanitary waste for babies, hospital, and other medical wastes, *waste from agriculture and industries*, organic household waste also contribute to the solid waste dumps.

Direct handling of solid waste can result in various types of infectious and chronic diseases with the waste workers and the rag pickers being the most vulnerable, along with the people living around the dumping sites. The waste contaminates the soil and water, which directly/ indirectly affects the general public as well.

5.4.1. Occupational Health and Safety

Occupational health refers to the identification and control of the risks arising from physical, chemical, and other workplace hazards in order to establish and maintain a safe and healthy working environment.

Occupational health and safety are concerned with the health, safety, and welfare of the people at work. It has a strong focus on primary prevention of hazards for the people who work there, the visitors and the general population. An occupational disease is a disease or disorder that is caused by the work or working conditions. Unsafe working environments impose a great risk of diseases and injuries to the people who work there.

The goals of occupational safety and health programs include providing a safe and healthy work environment not only to the employees but also to the family members, co-workers, visitors, customers, stakeholders, and also the general public by making sure the activities from workplaces or industries do not affect the environment. It includes the laws, standards, and programs that are aimed at making the workplace better for all.

The commonplaces of visit like shops, stadium, restaurants, playgrounds, streets, hotels, etc. have to guarantee safety of the patrons in terms of accident and disease prevention. To ensure this, occupational health and safety come into picture. It takes care of the safety issues and concerns related to physical hazards fire, light, sound radiation and extreme temperatures, accidents, ergonomic issues, chemicals, equipment safety, room conditions, and even psychological fallouts. Thus, occupational health and safety standards should be in place to mandate the removal, reduction, or replacement of workplace hazards.

As a consequence of occupational hazards, several diseases are known to occur, which ultimately affects the economic and social development. The occupational health risks are common in mines and industries. However, by carrying out routine factory inspection of all factory premises, the aim of improving the working conditions can be achieved. But these routine inspections are obstructed by the lack of financial resources and adequately trained personnel.

In case of inadequate inspections by Safety, Health, and Environment Officers in workplaces, there is a danger of polluting the environment and putting the health of the workers at risk. The Government needs to control and keep a constant check on the use of scarce resources so as to curb the pollution from industries and mines and work towards treatment of occupational diseases which have occurred dues to occupational hazards. The collaborative effort is required by Ministry of Labor and Ministry of Health to implement the Occupational Safety, Health, and Welfare Act of 1997 and thus make provision for the regulation of conditions of employment in workplaces with regard to safety, health, and welfare of employees. Also, now the act needs to revise in order to comply with the current and emerging challenges so as to reduce the impact of occupational health and safety issues on the health of people.

5.5. IMPACTS OF ENVIRONMENTAL DEGRADATION ON HUMAN HEALTH

Health of a population is influenced by many factors which include diet, sanitation, socio-economic status, literacy, and lifestyle. The rates of improvements in life expectancy, child survival, and adult survival more specifically in OECD regions improved significantly as there were changes in the above-mentioned factors during economic transitions.

OECD (The Organization for Economic Co-operation and Development) is an intergovernmental economic organization to stimulate economic progress and world trade. Better working conditions, increased GDP, and health expenditure per capita were the major reasons affecting life expectancy in OECD regions from the years 1970–1992.

However, this was the same period when air pollution started imposing its negative impacts on human health in OECD countries. In order to provide a complete picture of a population's health status, the various aspects which affect it can be combined in a measurement of the "burden of disease," as expressed for example in disability-adjusted life years (DALYs).

The 'DALYs' is a measure of overall disease burden, expressed as the number of years lost due to ill health, disability or early death. It is a way of comparing the overall health and life expectancy of different countries. They give an indication of how the duration of disease combined with the impact of disease can alter the ability of people to live normal lives as compared to those with no disease (Murray and Lopez, 1996).

Daly's approach is used to calculate the average burden of disease for all OECD countries, for OECD countries grouped by income level, and for non-OECD countries. The findings clearly suggest that with regard to both total burden of disease and the health conditions related to environmental degradation, there is a difference between OECD countries and non-OECD countries. Income is the basis of the environment-related share of the burden of disease like with higher environmental shares generally occurring in lower-income countries. In OECD countries, this share is estimated to be 2–6% of the total burden of disease.

In OECD countries, health is lost primarily through the non-communicable (chronic, degenerative) diseases while in non-OECD countries, the majority of the burden of disease can be attributed to communicable disorders (e.g., infectious, maternal, prenatal). In OECD countries, conditions like heart disease and depression make up a major portion of the burden of disease.

In OECD countries, the percentage of young children suffering from disease is quite lower (7% of the total burden of disease) than the non-OECD countries where diseases in children under four years old account for 50% of the total burden of disease. In non-OECD countries, the major factors which determine the environmental share of health problems majorly include those related to poverty like limited access to proper food, housing, health care, and drinking water. While in OECD countries, the environmental determinants of human health are related more to the exposure to air pollutants (particularly in urban areas) and chemicals in the environment than to poor living conditions.

Dioxins are a group of chemically-related compounds that are persistent environmental pollutants (POPs).

Other human health risks that have recently received considerable attention include animal feed contaminants and their negative effects on both animal and human health. Unsafe livestock feeding practices and their contamination by persistent organic pollutants, toxic metals and natural toxins is increasing the risk of chemical and microbiological contaminants being transferred into food-producing animals, and thus the toxins reach the food chain unintentionally.

Poultry feed accidentally contaminated by **dioxins** can move up the food chain to humans. Similarly, "mad cow disease" (BSE) can be caused in livestock by using feed contaminated by diseased animal remains. The human form of mad cow disease is called variant Creutzfeldt-Jakob disease (CJD), which is fatal. This can happen if the consumer eats nerve tissue (the brain and spinal cord) of cattle that were infected with mad cow disease.

Exposure to chemicals and air pollutants can affect health in varying manner ranging from just allergies to the life-threatening cancers. Urban air pollutants can cause asthma, allergies, respiratory diseases at low exposure levels and cardiovascular diseases if the exposure is continuous or long-term. Neurological disorders and various cancers have been reported due to exposure to heavy metals. POPs can also cause various cancers and are suspected of causing birth defects and reproductive disorders. In OECD countries, the environmental threats related to human health

Air pollution and exposure to hazardous chemicals are important causes of the environment-related threats to human health in OECD countries. They have significant impact on human health. A well-known example is the effect on the ozone layer of ozone-depleting substances (ODS) used in cooling systems and spray cans.

The depletion of the ozone layer has led to increased exposure to UV-radiation and a greater risk of skin cancer. Many OECD countries have shown concern, and the production of ODS has considerably decreased in recent years, still the exposure levels to UV- radiation are still above acceptable levels in many regions of the world.

5.6. THE HOME ENVIRONMENT PROBLEMS

Managing health care and environmental problems at home and neighborhoods is a challenge confronted by many people and tend to have immediate and health-threatening effects. The most affected include older persons, children, and younger adults with disabilities, who are especially vulnerable to their environments. People with low incomes, multiple chronic illnesses, limited social supports, and health disparities, as well as those who live in unsafe neighborhoods or poor-quality housing, are also at great risk.

Women too are at enormous risk and more susceptible to infections as they carry a disproportionately high responsibility for household chores, typically spend more time in and around the home than men, and often also have their economic activities located within the home area. They tend to be affected directly as they are also responsible for caring for the sick and maintaining cleanliness in the home. Women also have a higher stake in the home and neighborhood environments and tend to be more active in organizing neighborhood improvements. People's living environments are where improvements can do the most to enhance human health in developing countries. A majority of vulnerable individuals do not receive effective health care for their chronic illnesses and live in housing that does not support their independence or functioning.

Poor water, sanitation, and hygiene generate multifaceted health risks. When water supplies and sanitary conditions are inadequate, they constitute some of the major health problems in the developing world and pose threat of various diarrheal, and other diseases are spread via fecal-oral routes. Poor planned or unplanned housing and food systems, along with other social and lifestyle factors, are drivers in the epidemic of noncommunicable diseases, which are linked to risks and hazards such as air pollution, poor diet, and domestic injury.

5.7. ROLE OF THE HEALTH SECTOR

The health sector is defined as the set of values, standards, institutions designed to prevent and control disease, care for the ill, and conduct health research and training. In the recent years, comprehensive health and environmental policies and strategies have been developed, and much progress has been made in this direction, but still, in many countries of the world, the development of such health-related sectors has been really slow. This has been due partly to the fact that there are many gaps in knowledge and perceptions of insufficient evidence on which to act. Besides this, it is also due to the other real challenges to the health sector of addressing policy needs with respect to new and expanded areas such as energy, **agriculture**, industrialization, and advanced technology. Therefore, the health sector plays a key role in collaborating with other sectors and organizations in formulation of disaster risk reduction policies with specific short-, medium-, and long-term objectives and strategies to achieve the objectives efficiently and effectively, and thus contributing to health protection and promotion. While establishing the health care sector plan, system must be set up to establish clear links between the functions, roles, and responsibilities of all health sector institutions and levels, including each division and department in the ministry of health.

Agriculture is the process of producing food, feed, fiber and many other desired products by the cultivation of certain plants and the raising of domesticated animals (livestock).

With regard to the sustainable development, Agenda 21 presents the global action plan which aims to meets the needs of the present generation without harming the ability of future generations to meet their needs. This comprehensive plan gives the health authorities an opportunity to monitor and report on implementation of the policies and strategies at local, national, regional, and international levels and make efforts to reverse the trend of environmentally damaging and health-threatening development. The Earth Summit, held in 1997 inspired a number of countries to take initiatives and include a stronger health focus in national planning for sustainable development. The health care systems in recent times are beginning to wake up to the best scenario of more sustainable health systems which are much cheaper, provide better care, and are better for the environment. With the emergence of such complex environmental and health systems, it has become necessary to define more clearly the responsibility of the health sector. This ensures that the activities of all related sectors and organizations contribute positively to health protection and promotion.

REVIEW QUESTIONS

1. How urbanization and industrialization are aiding in pollution?
2. Explain some environmental pollutants.
3. What are greenhouse gases?
4. What is the relation between health and environmental pollution?
5. Define "source-effect chain."
6. Explain environmental-related health issues.
7. What do you mean by urban sanitation?
8. What are the impacts of environmental degradation on human health?
9. Why solid waste management is necessary to get rid of pollution?
10. What are various risks and challenges related to environmental pollution?

REFERENCES

1. Briggs, D., (2003). Environmental pollution and the global burden of disease. *British Medical Bulletin*, *68*(1), 1–24. doi: 10.1093/bmb/ldg019 (Accessed on 18 June 2019).
2. *Health & Environment Tools for Effective Decision-Making*, (2004). Retrieved from: https://www.who.int/heli/publications/helirevbrochure.pdf (Accessed on 18 June 2019).
3. *Issues in Health, Environment and Sustainable Development: An Overview*, [eBook] (p. 10). Retrieved from: https://www.who.int/mediacentre/events/IndicatorsChapter1.pdf (Accessed on 18 June 2019).
4. Kelishadi, R., (2012). *Environmental Pollution: Health Effects and Operational Implications for Pollutants Removal*. Retrieved from: https://www.hindawi.com/journals/jeph/2012/341637/ (Accessed on 18 June 2019).
5. Kjellén, M., (2001). *Health and Environment* [eBook] (p. 58). Swedish International Development Cooperation Agency. Retrieved from: https://www.sida.se/contentassets/f878f9d4943e4e1c97f6d6f62edff9d2/20012-health-and-environment.-issue-paper_641.pdf (Accessed on 18 June 2019).
6. Kumwenda, S., Samanyika, Y., Chingaipe, E., Mamba, K., Lungu, K., Mwendera, C., & Kalulu, K., (2013). *The Emerging Environmental Health Risks and Challenges for Tomorrow: Prospects for Malawi* [eBook] (p. 10). Retrieved from: https://www.ifeh.org/wehd/2013/Article_WEHD_Malawi.pdf (Accessed on 18 June 2019).
7. OECD Publications, (2001). *Human Health and the Environment* [eBook] (p. 7). Retrieved from: http://www.oecd.org/health/health-systems/32006565.pdf (Accessed on 18 June 2019).
8. *The Link Between the Environment and Our Health*. Retrieved from: https://www.scientificamerican.com/article/environment-and-our-health/ (Accessed on 18 June 2019).
9. The WHO-UNEP Health and Environment Linkages Initiative (HELI) https://www.who.int/heli/publications/helirevbrochure.pdf

Chapter 6

ENERGY USE AND ITS ENVIRONMENTAL IMPLICATIONS

LEARNING OBJECTIVES:

In this chapter, you will learn about:

- The use of energy and its resources.
- Environmental impact of energy utilization.
- Trends of energy consumption.
- Environmental issues linked with non-conventional energy.
- Environmental concerns related to the biomass use.

Key Terms:

- Biomass innovations
- Gaseous fuels
- Geothermal plants
- Mechanical squanders
- Petroleum based fluid
- Photovoltaic frameworks
- Renewable power generators
- Ventilation

6.1. INTRODUCTION

Worldwide, around 40% of the complete world yearly energy utilization is done by buildings. The vast majority of this energy is for the arrangement of heating, air conditioning, lighting, and cooling. Moreover, expanding attention to the natural effect of carbon dioxide and nitrogen outflows and chlorofluorocarbons (CFCs) sets off a restored enthusiasm for environmentally friendly heating and cooling innovations under the 1997 Montreal Protocol. Therefore, governments consented to eliminate synthetic substances utilized as refrigerants that can possibly demolish stratospheric ozone. It was along these lines thought about to lessen energy utilization and reduction the rate of exhaustion of world energy reserves and contamination of the earth.

On the other hand, one method for diminishing building energy utilization is to plan buildings that are increasingly practical in their utilization of energy for hot water supply, ventilation, heating, lighting, and cooling. Moreover, compliant measures, especially hybrid or natural **ventilation** as opposed to air conditioning, can significantly lessen essential energy utilization. Be that as it may, misuse of sustainable energy in agricultural greenhouses and buildings can likewise essentially contribute toward diminishing reliance on fossil fuels. Hence, advancing innovative renewable applications and fortifying the renewable energy market will add to the conservation of the biological system by lessening emissions at global and local levels. This will likewise add to the enhancement of ecological conditions by supplanting conventional fuels with renewable energies that produce no air contamination or ozone-harming gases. Finally, the arrangement of good indoor environmental quality (IEQ) while accomplishing energy and cost proficient task of the ventilating, air conditioning, and heating plants in building represents a multivariate issue. The solace of occupants of the building is subject to numerous ecological parameters including temperature, airspeed, relative dampness, and quality along with noise and lighting.

On the other hand, the general goal is to give a higher level of building performance, which can be characterized as IEQ, energy efficiency (EE), and cost efficiency (CE). IEQ is the apparent state of solace that building inhabitants experience due to the psychological and physical conditions to which they are exposed by their environment.

Ventilation is the intentional introduction of outdoor air into a space and is mainly used to control indoor air quality by diluting and displacing indoor pollutants; it can also be used for purposes of thermal comfort or dehumidification.

The primary physical parameters influencing IEQ are quality, relative humidity, temperature, and speed. EE is identified with the arrangement of the ideal natural conditions while expending the negligible amount of energy. CE is the monetary use on energy in respect to the dimension of natural solace and efficiency that the structure inhabitants achieved. Moreover, the overall CE can be upgraded by improving the IEQ and the EE of a structure. The expanded accessibility of dependable and effective energy services stimulates new improvement choices.

However, at present, renewable energy gives just a minor portion of its potential electricity yield around the world. In any case, various investigations have more than once revealed that renewable energy can be quickly deployed to give a huge share of future power needs, even in the wake of representing potential imperatives. For example, REN21 Renewables 2010 Global Status Report renewable energy source replaces traditional fuel in four different zones: hot water or space heating, motor fuels, electricity generation, and rural (off-grid) energy services:

- **Power Generation:** Renewable energy gives 19% of power generation around the world. Renewable power generators are spread crosswise over numerous nations, and wind power alone as of now gives a huge share of power in certain zones: for instance, 40% in the northern German territory of Schleswig-Holstein, 49% in Denmark and 14% in the U.S. state of Iowa.

> Few nations get the vast majority of their capacity from renewables, including Norway (98%), Austria (62%), Iceland (100%), New Zealand (65%), Sweden (54%) and Brazil (86%).

- **Heating:** Sun-powered high temp water makes an essential contribution to renewable heat in numerous nations, most prominently in China, which presently has 70% of the worldwide aggregate (180 GWh). The greater part of these frameworks is introduced on multifamily apartment structures and meets a bit of the high-temperature water needs of an expected 50 to 60 million families in China.

Moreover, around the world, the total number of solar waters warming frameworks meets a part of the water warming needs of more than 70 million family units. The utilization of biomass for warming keeps on developing also. In Sweden, national utilization of biomass energy has outperformed that of oil. Direct geothermal for warming is additionally developing rapidly.

- **Transport Fuels:** Inexhaustible biofuels have added to a huge decrease in oil utilization in the United States since 2006. The 93 billion liters of biofuels generated globally in 2009 uprooted what might be compared to an expected 68 billion liters of fuel, equivalent to about 5% of world gas generation.

6.2. ENERGY CONSUMPTION TRENDS

In the year 2008, 81% of our energy supply was from petroleum products (26.8% coal, 33.5% oil, and 20.8% gas, separately) while 12.9% was sorted as "other" (solar, wind, peat, hydro biofuels, geothermal power, and so on.), and 5.8% was nuclear. Oil was the most well-known energy fuel, and coal and oil jointly represented over 60% of the world energy supply in 2008. Starting at 2010, utilization of fossil fuel as an energy source involved over 80% of total energy devoured.

The industrial revolution was fueled by the coal in the eighteenth and nineteenth centuries. In any case, with the arrival of the airplanes, automobiles, and the developing utilization of power among buyers, oil turned into the prevailing fuel amid the twentieth century. The development of oil as the biggest non-renewable energy source was additionally empowered by relentlessly dropping costs from 1920 until 1973.

Nonetheless, after the oil stuns of 1973 and 1979, amid which the cost of oil increased from $5 to $45 per barrel, there was a move far from this specific asset. These days, natural gas, nuclear power, and coal are the most mainstream fuels for **electricity generation**. However, recent preservation measures have incredibly expanded energy productivity. In the United States, the average vehicle has dramatically increased the quantity of miles per gallon. Japan, which endured the worst part of the oil stuns, has additionally made dynamite upgrades to its innovation. Presently Japan has the most

Electricity generation is the process of generating electric power from sources of primary energy.

noteworthy energy productivity on the planet.

6.2.1. Nuclear Power

As of December 2009, the world had 436 nuclear reactors. Since commercial nuclear energy began in the mid-1950s, 2008 was the first year that no new nuclear power plant was connected to the grid, although two were connected in 2009. Annual generation of nuclear power has been on a slight downward trend since 2007, decreasing 1.8% in 2009 with nuclear power still meeting 13–14% of the world's electricity demand. Moreover, consumption of fossil fuel resources has led to global warming and climate change. In most parts of the world, little effort is being made to slow these changes. If we reach (or have already reached) a maximum rate of petroleum extraction, then fossil fuels will no longer be a viable source of energy. However, if we explore viable alternative energy resources, we could reduce our impact on the environment. Thus, emerging technologies could lead to more efficient energy generation, which could temper environmental damage. The use of these new methods incorporates practices that are informed by new paradigms such as systems ecology and industrial ecology.

Greenhouse gas is a gas that absorbs and emits radiant energy within the thermal infrared range.

ON the other hand, global warming and climate change are generally accepted as being caused by anthropogenic (man-made) **greenhouse gas** (GHG) emissions. The majority of GHG emissions are due to burning fossil fuels, while some is due to deforestation. As of December 2009, the world had 436 atomic reactors. However, since business atomic energy started in the mid-1950s, 2008 was the principal year that no new atomic power plant was associated with the network; however, two were associated in 2009. Yearly generation of atomic power has been on a slight descending pattern since 2007, diminishing 1.8% in 2009 with atomic power as yet meeting 13 to 14% of the world's demand for electricity.

Arriba Thus, the utilization of fossil fuel resource has prompted climate change and global warming. In many pieces of the world, little exertion is being made to moderate these changes. On the off chance that we reach (or have already reached) a most extreme rate of oil extraction, at that point fossil fuel will never again be a practical wellspring of energy. Notwithstanding, in the event that we investigate viable elective energy assets, we could lessen our effect on the earth.

Nevertheless, recently rising technologies could prompt increasingly effective energy generation, which could temper environmental damage. However, the utilization of these new strategies joins practices that are informed by new ideal models, for example, industrial ecology and systems ecology. Thus, climate change and global warming are commonly acknowledged as being brought about by anthropogenic (man-made) ozone-harming gas emanations. Most of the ozone-harming gases outflows occurs because of the consumption of non-renewable energy sources, while some are because of deforestation.

6.3. COMBUSTION AND THE ENVIRONMENT

Contaminant content and physical forms are the two attributes of power that most influence their poison emanations when burned. It is thus, commonly hard to pre-blend solid fuels adequately with air to guarantee great burning in basic small-scale gadgets, for example, family stoves. Subsequently, despite the fact that most biomass fuels contain a couple of poisonous contaminants, they are normally not burned completely in family unit stoves, thus produce a wide range of wellbeing harming contaminations (Figure 6.1).

Figure 6.1: Combustion and environment.

Source: https://media.defense.gov/2012/Nov/17/2000095670/-1/-1/0/121111-F-ZZ999-641.JPG

However, wood and different biomass fuels would generate minimal other than carbon dioxide, water, and non-lethal items, when combusted totally. Be that as it may, practically speaking now and then as much as one-fifth of the fuel carbon is diverted to results of inadequate ignition, huge numbers of which are imperative wellbeing harming toxins.

Coal, then again, isn't just hard to burn totally on the grounds that it is strong, yet additionally regularly contains critical inborn contaminants that add to its emanations of wellbeing harming contaminations. Most noticeable among such discharges are sulfur oxides.

In any case, in numerous territories, coal additionally contains fluorine, arsenic, and other poisonous components that can prompt genuine wellbeing harming contaminants. A huge number of individuals in China, for instance, are exposed to such toxins from coal used in households. **Gaseous fuels** and Petroleum based fluid, for example, liquefied petroleum gas and kerosene, can likewise contain sulfur and different pollutants, however in a lot of smaller sums than in numerous coals.

Gaseous fuels are hydrocarbons, hydrogen and carbon monoxide mixtures present in gaseous state which forms the basis of potential heat energy or light energy that can be readily disseminated by means of pipes from the origin to the place of consumption.

Also, their physical structures permit much better pre-blending with air in simple gadgets. Thus, guaranteeing a lot of higher ignition efficiencies and lower outflows of wellbeing harming toxins as results of fragmented burning. Moreover, stoves for gaseous fuels and petroleum-based fluid are significantly more energy effective than those for coal. Accordingly, emanations of wellbeing harming contaminants per meal from these fuels are something like an order of extent not exactly those from solid fuel.

However, in order to gauge the wellbeing harm from contamination, it is important to consider the measure of contamination discharged. Similarly, vital, nonetheless, is the conduct of the populace in danger. Indeed, even a lot of contamination won't have much wellbeing sway if little of it contacts individuals. In any case, a generally small measure of contamination can have a major impact on wellbeing if it is discharged at the places and occasions where individuals are available, such that a huge amount is taken in.

In this manner, it is important to look at where the contamination is as well as at where the general population is located. Sadly, contamination from family unit stoves is discharged right at the places and occasions where individuals are available, that is, in each family unit consistently. This is the recipe for high contamination exposures: huge measures of contamination frequently discharged in ineffectively ventilated spaces at simply the occasions when individuals are available.

In addition, as a result of their almost all-inclusive duty regarding cooking, ladies, and their most youthful youngsters are commonly the most exposed. Accordingly, in spite of the fact that the aggregate sum of wellbeing harming contamination discharged from stoves worldwide isn't high with respect to that from huge scale utilization of fossil fuels, human exposures to various imperative contaminants are a lot bigger than those generated by open air contamination. Subsequently, the wellbeing impacts can also be expected to be higher.

6.4. ENVIRONMENTAL ISSUE RELATED TO NON-CONVENTIONAL ENERGY

Non-conventional wellspring of energy generation has some pragmatic constraints.

Some broad disadvantages are: trouble in creating large amount of power in centralized way as in the event of fossil fuels, reliance on the climate conditions at the time and in the locale of utilization, generally costly to procure and set up the types of gear essential for the producing power and specialized limitations in (effectiveness) of creating power from the non-conventional resource.

Thus, other than some innovative and monetary imperatives, it additionally has some related ecological issues which are talked about beneath:

6.4.1. Wind Energy

Wind energy is viewed as one of the cleanest and most manageable approaches to create power as it delivers no harmful contamination or GHGs. The wind is plentiful and inexhaustible, which can make it a practical and vast scale option in contrast to petroleum products.

Figure 6.2: Wind energy.

Source: https://www.maxpixel.net/static/photo/1x/Windraeder-Wind-Power-Energy-Wind-Energy-3388600.jpg

In spite of its enormous potential, there are sure ecological effects related with wind power generation that ought to be mitigated and recognized (Figure 6.2).

6.4.1.1. Environmental Impacts of Wind Power

Harnessing power from the wind is one of the cleanest and most maintainable approaches to create power as it delivers no harmful contamination or GHGs outflows. Wind is inexhaustible, bounteous, which makes it a suitable and substantial scale option in contrast to non-renewable energy sources. In spite of its huge potential, assortments of natural effects related with wind power generation that ought to be mitigated and recognized.

6.4.1.2. Land Use

Wind turbines put in level regions that ordinarily utilize more land than those situated in bumpy territories. It has been assessed that around one acre of land per MW is disturbed for all time and under 3.5 acres of land/MW are aggravated briefly amid development. Further, the rest of the land can be utilized for an assortment of other gainful purposes, including construction of highways, animal grazing, hiking trails, and agribusiness.

Arriba Therefore, wind facilities can be introduced close brownfields (relinquished or underused industrial land), which essentially diminishes worries about land use. Thus, offshore wind facilities, which require bigger measure of space in light of the fact that the turbines and sharp edges are greater than their territory-based partners. Arriba It likewise blocks an assortment of other marine exercises, for example, fishing, oil, and gas extraction, recreational exercises, navigation, sand, and rock extraction, and **aquaculture**. However, utilizing best practices in arranging and siting can help in limiting potential land use impacts.

Aquaculture is the farming of fish, crustaceans, molluscs, aquatic plants, algae, and other organisms.

6.4.1.3. Wildlife

The effect of wind turbines on bats and winged animals and wildlife, has been broadly reported and concentrated by specialists. Arriba Moreover, specialists, discovered proof of feathered creature and bat deaths from crash with wind turbines and because of changes in pneumatic stress brought about by the turning turbines, just as because of natural surroundings obliteration. The National Wildlife Conservation Council (NWCC) presumed that these effects are moderately low and don't represent any danger to species populaces.

6.4.2. Solar Energy

Like the wind, the sun is additionally an enormous source for creating feasible and clean power without lethal contamination or global warming outflows. Thus, the natural effects related with solar power can incorporate habitat loss and land use, water use, and utilization of unsafe materials in manufacturing. Moreover, the sorts of effects fluctuate enormously relying upon the size of the framework and the innovation utilized – photovoltaic (PV) sun-powered cells or concentrating solar thermal plants (CSP) (Figure 6.3).

Figure 6.3: Solar energy.

Source: https://images.pexels.com/photos/356036/pexels-photo-356036.jpeg?cs=srgb&dl=alternative-energy-building-clouds-356036.jpg&fm=jpg

6.4.2.1. Land Use

Contingent upon their area, sunlight-based offices can raise issues about habitat loss and land degradation. Land region prerequisite thus varies upon the adopted innovation, the geography of the territory, and the force of the sun-oriented radiation. Moreover, evaluations for utility-scale PV frameworks go from 3.5 to 10 acres per MW, while gauges for CSP facilities are somewhere in the range of 4 and 16.5 acres of land per megawatt.

There are less opportunities for sun-oriented undertakings to share the land with other land use movement. Appropriate siting at lower quality areas, for example, brownfields, surrendered mining area, waterway, and transmission hallways, small scale sun-based PV exhibits based on homes or business structures, help to limiting the land use sway.

6.4.2.2. *Water Use*

Sun oriented PV cells don't utilize water for producing power. Nonetheless, concentrating sunlight based warm plants, require water for a cool down. Thus, water use relies upon the plant area, plant structure, and the kind of cooling framework to acquire. However, concentrating sun based warm plants that utilization wet recycling innovation with cooling towers pull back somewhere in the range of 600 and 650 gallons of water for every MW-Hr. of power generated.

On the other hand, CSP with dry cooling innovation can decrease water utilization by around 90% yet at staggering expense and essentially less compelling at temperatures over 100°F. Unexpectedly, territories with most noteworthy potential for sunlight-based energy likewise frequently, in general, tend to be the dry and water shortage zones; thus, a cautious thought of these water **tradeoffs** is fundamental.

Tradeoff is a situational decision that involves diminishing or losing one quality, quantity or property of a set or design in return for gains in other aspects.

6.4.3. Geothermal Energy

Geothermal power plants (also known as hydrothermal plants) are situated close to geologic 'hot spots' where hot liquid rock is near to the crust of the earth and creates high-temperature water or different districts of hot dry rock.

Geothermal plants vary as far as the innovation they use to change over the asset to power and the kind of cooling innovation they acquire (water cooled and air cooled). Natural effects shifts relying upon the change and cooling innovation acquired (Figure 6.4).

Figure 6.4: Geothermal energy.

Source: https://cdn.pixabay.com/photo/2016/02/19/07/43/hole-1208874_960_720.jpg

6.4.3.1. Water Quality and Use

Geothermal power plants can have impacts on water quality. Heated water is taken from underground repositories frequently contains abnormal amounts of salt, sulfur, and various other minerals. Moreover, water is utilized by geothermal plants for re-injecting and cooling. Contingent upon the cooling innovation utilized, geothermal plants additionally require 1,700 to 4,000 gallons of water for each MW-Hr.

Liquid is a nearly incompressible fluid that conforms to the shape of its container but retains a (nearly) constant volume independent of pressure.

Be that as it may, the utilization of geothermal **liquids** and closed loop innovation (reinjection of same water again into the store) can lessen the overall demand for freshwater as well as safeguards contamination and land subsidence.

6.4.3.2. Air Emissions

In closed loop frameworks, air emanations are insignificant. Be that as it may, the open loop frameworks produce boron, H_2S, NH_3, CO_2, and

CH_4. H_2S, which has a particular 'spoiled egg' smell, is the most widely recognized emanation. Its outcomes can be seen in the form of damaged crops, forests, acid rain, and acidification of water body, health problems, and soils.

Arriba Furthermore, although, SO_2 outflows from geothermal plants are roughly thirty times lower per MW-Hr. than from coal plants, small measures of mercury emanations from the geothermal plant can be alleviated utilizing mercury channel innovation. However, scrubbers can decrease air discharges, transfer of muck at unsafe waste sites.

6.4.3.3. Land Use

The measure of land required by a geothermal plant differs relying upon the type of energy conversion system, the type of cooling system, the arrangement of wells and piping systems, the properties of the resource reservoir, auxiliary building and the substation needs, and the amount of power capacity. Around 13 acres of land is required for the generation of 1MW power.

However, numerous geothermal sites are situated in sensitive and remote environmental zones, so venture organizers must consider in their arranging procedures to keep away from the danger of land subsidence. Aqueous plants are sited on land "problem areas," which will, in general, have more elevated amounts of risks of the earthquake. Tremor chance related with upgraded geothermal frameworks can be limited by siting plants a suitable distance far from significant fault points.

6.4.3.4. Small Hydropower

Hydroelectric power incorporates both small run-of-the-river plants and enormous hydroelectric dams; although, the advancement of vast hydropower turned out to be progressively problematic due to socio-ecological concerns. Additionally, small scale hydro improvement has not met desires either. The ecological issues as impact to fish living space would constrain small hydropower advancement in the nation. Moreover, the effect of vast dams is greater to the point that there is pretty much every time clashes exist with the improvement of such ventures. The reserves made by such activities as often as possible immerse/submerge expansive territories of farmland, woods, scenic areas, untamed life natural surroundings, grand regions, and furthermore towns. However, the dams can cause noteworthy changes in stream biological systems, including aquatic habitat both downstream and upstream.

As a result, little hydropower plants utilizing reserves can cause comparative sorts of harm, yet on a little scale. With expanding progressively stringent natural guidelines, the advancement of cost-effective alleviation measures is essential. Truth be told, **hydropower** is moving toward the breaking point of its potential and its commitment to add up to power generation may decrease in future because of expanded demand of water assets for drinking reason and farming, declining precipitation, and endeavors to ensure jeopardized fish and natural life.

Hydropower is power derived from the energy of falling or fast-running water, which may be harnessed for useful purposes.

6.5. ENVIRONMENTAL IMPLICATIONS OF INCREASED BIOMASS USE

Biomass use is relied upon to increment with evolving financial aspects. It is assessed that 14.8 quads of biofuel could be utilized constantly by the year 2030 by applying quickened research, advancement, and exhibit to productive innovations. It would require rigorous management of timber and 192 million acres of land of unused cropland, principally in the South, Southeast, and Midwestern United States. This speaks of an expansion of roughly three times of current biomass use.

Biomass fuels, later on, will be like those utilized today. Woody powers from cropland and forest would progressively supply electric and industrial applications. Wood buildup powers will keep on breaking down in quality as wood is reallocated to developing fiber marketplaces. Buildup supply will be enhanced by expanded woods collecting and ranch breeds.

Farming fuels will be utilized whenever justified by neighborhood financial matters and mechanical advances. Urban wood utilization will proceed. Fuels for transportation will be generated from biomass plants. New assets, enough to generate in excess of 10 quads, will get from intensive cultivation of cropland and escalated management of timberland. Air contamination from private wood ignition or harvesting on natural surroundings, attract consideration regarding the potential expenses of biomass vitality. The bioenergy network must address these vast scale ecological concerns before pressure from the public limit the utilization of biomass innovations. However, projects are now and again contradicted in light of issues that the natural expenses of biomass transformation to vitality might be critical. It is thus, a motivation behind this review to distinguish which of these territories of natural concern have been sufficiently tended to and what regions still need study or illumination.

6.5.1. The Environmental Concerns

Globally, concerns for the earth have created an abundance of writing on the improvement and utilization of biomass for vitality. Numerous parts of biomass generation and transformation have been contemplated, and some have been convincingly tended to or are in effect right now tended to by numerous associations. A few ecological investigations show up in the task records and distributions of state and neighborhood governments, which are some of the time supported by DOE provincial projects.

Moreover, somewhere a long time ago around, two districts United States have arranged nonexclusive outlines of ecological effects for biomass vitality. California in arranged a presentation of negative natural effect for various biomass vitality showing ventures and the Tennessee Valley Authority (TVA) assessed the territorial ecological effect of biomass transformation offices. However, generic effect explanations help to support the area of biomass facilities in a specific zone and in order to offer approach options intended to empower use of **biomass**, the motivation behind an environmental study, announcement of negative ecological effect, or natural effect proclamation is to distinguish those components fitting to a particular site instead of to an innovation.

Biomass is plant or animal material used for energy production (electricity or heat), or in various industrial processes as raw material for a range of products.

6.5.2. Biomass versus Fossil Fuels

As a sustainable asset, biomass generation, contrasted with non-renewable energy sources, benefits nature, especially its capability to keep up or re-establish the earth and to relieve worldwide environmental change. Biomass consuming, for instance, is a moderate contributor to visibility, ultraviolet absorption, and radiation, however, it is not a controlling element of real significance like coal burning or petroleum. Thus, it doesn't add to the issues of oxidation, acidity, or oxidants to the level of the petroleum derivatives. In the US since the 1990 Clean Air Act perceives the advantages of biomass ignition by giving motivating forces to coal-consuming utilities to utilize biomass. Today in 2019 the US government refuses to sign

6.5.3. Environmental Impacts of Biomass

Biomass power plants share a few similitudes with petroleum product power plants: both include the ignition of a feedstock to create power. Hence, biomass plants raise comparative, yet not indistinguishable, issues about air emanations and water use as petroleum derivative plants. Biomass control plants, similar to coal and flammable gas terminated power plants, require water for a cool down. However, land use impacts from biomass power creation are driven basically by the sort of feedstock: either a waste stream or a vitality crop that is developed explicitly for producing power. There are an Earth-wide temperature boost discharges related with developing and collecting biomass feedstock, transferring feedstock to the power plant, and igniting or gasifying the feedstock.

On the other hand, combustion and transportation outflows are generally equal for a wide range of biomass. Be that as it may: an unnatural weather change emanation from the sourcing of biomass feedstock shift varies vastly. It was once generally imagined that biomass had net zero an unnatural weather change discharges, in light of the fact that the developing biomass assimilated an equivalent measure of carbon as the sum discharged through burning, however, at this point it is comprehended that some biomass feedstock sources are related with significant an Earth-wide temperature boost outflows.

Hence, helpful biomass assets incorporate vitality crops that don't contend with nourishment crops for land, bits of yield buildups, for example, corn stove or wheat straw, economically gathered wood and timberland deposits, and clean metropolitan and mechanical squanders.

6.6. ENVIRONMENTAL IMPACTS OF GEOTHERMAL ENERGY

Geothermal power is a generally kindhearted wellspring of vitality. Generally, the effects of improvement are sure. Worldwide geothermal vitality use builds yearly since it is an alluring option in contrast to consuming imported and residential petroleum products. Power generation from geothermal assets includes much lower GHGs emanation rates than that from petroleum products. As per the International Atomic Energy Agency (IAEA), supplanting one kilowatt-hour (kWh) of fossil power with a kWh of geothermal power it diminishes the assessed impact of global warming by around 95%.

Notwithstanding, geothermal advancement could have few negative effects if suitable relief activities and checking plants are not set up. Any extensive scale development and penetrating activity will deliver visual effects on the scene, make clamor and squanders and influence nearby economies. A few nations have exacting ecological guidelines with regards to a portion of the effects related with geothermal improvement, and others do not.

Ecological issues generally tended to improve with the advancement of geothermal fields that incorporate waste disposal, air quality, water quality, land use issues, biological resources, geologic risks, and noise. The conservation of **groundwater** is significant amid the penetrating stage. However, the groundwater is to be overseen reasonably because it is a piece of the biological system, is a living space for creatures and plants, and has a significant role in the daily lives of neighborhood occupants.

Groundwater is the water present beneath Earth's surface in soil pore spaces and in the fractures of rock formations.

On the other hand, the primary visual effect amid the development stage is the nearness of a boring apparatus, yet once a task is in the generation stage the apparatus isn't required, and the vitality center impression is extremely little. Due to low outflows, the geothermal power plants likewise meet the most stringent clean air benchmarks. It ought to be noticed that every single geothermal plant needs to meet different local and national ecological principles and guidelines, in spite of the fact that discharges are not routinely estimated underneath a specific limit, and outflows from geothermal plants ordinarily fall beneath this edge.

However, the rundown of hindrances coming about because of ecological guidelines can be somewhat long. Ecological guidelines ought to incorporate groundwater assurance incl., weight issues, soil security yet additionally convention on miniaturized scale seismicity, and surface issues. Thus, for work security, development, and traffic, any enactment appropriate for comparable exercises in mining, boring, development, and so forth ought to be applied.

REVIEW QUESTIONS

1. Explain the trends of energy consumption.
2. How the environment is affected by combustion?
3. Define biomass.
4. What do you mean by conventional and non-conventional energy?
5. Describe environmental issue related to non-conventional energy.
6. Define wind energy, solar energy, and geothermal energy.
7. Compare the effects of biomass and fossil fuels.
8. How geothermal energy impacts the environment?
9. Illustrate those four key areas in which renewable energy replaces conventional fuels.
10. What are the impacts of wind energy on wildlife?

REFERENCES

1. *Energy Use and the Environment | Boundless Chemistry*. Retrieved from: https://courses.lumenlearning.com/boundless-chemistry/chapter/energy-use-and-the-environment/ (Accessed on 18 June 2019).
2. Holdren, J., & Smith, K. (1999) *Energy, the Environment, and Health* [eBook] (p. 50). Retrieved from: http://user.iiasa.ac.at/~gruebler/Lectures/Graz-04/wea_chapter3.pdf (Accessed on 18 June 2019).
3. Klugmann-Radziemska, E., (2014). *Environmental Impacts of Renewable Energy Technologies* [eBook] (p. 6). Retrieved from: http://www.ipcbee.com/vol69/021-ICEST2014-A1026.pdf (Accessed on 18 June 2019).
4. Miles, T., (1992). *Environmental Implications of Increased Biomass Energy Use* [eBook] (p. 56). National renewable energy laboratory. Retrieved from: https://www.nrel.gov/docs/legosti/old/4633.pdf (Accessed on 18 June 2019).
5. Omer, A., (2009). Energy use and environmental impacts: A general review. *Journal of Renewable and Sustainable Energy*, *1*(5), 053101. doi: 10.1063/1.3220701 (Accessed on 18 June 2019).
6. Sharma, A., & Rout, C. (2017) Potential of non-conventional source of energy and associated environmental issue: An Indian scenario [eBook] (p. 5). *International Journal of Computer Applications*. Retrieved from: https://pdfs.semanticscholar.org/4943/cda1f0d69ebe867ef4bd235d85dedc42ced8.pdf (Accessed on 18 June 2019).
7. Tsoutsos, T., Frantzeskaki, N., & Gekas, V., (2003). *Environmental Impacts from the Solar Energy Technologies* [eBook] (p. 8). Elsevier. Retrieved from: https://www.circleofblue.org/wp-content/uploads/2010/08/Tsoutsos_Frantzeskaki_2006_EIA_ST.pdf (Accessed on 18 June 2019).

Chapter 7

SOIL SCIENCE AND THE ENVIRONMENT

LEARNING OBJECTIVES:

In this chapter, you will learn about:

- Soil and its fundamental properties.
- Contamination of soil by various parameters.
- Natural and anthropogenic activities leading to soil degradation.
- Process of soil erosion and its impact on the environment.

Key Terms:

- Acid rain
- Cash crops
- Contamination
- Deforestation
- Heavy metals
- Industrialization
- Radionuclides
- Sea-salt sprays
- Soil erosion
- Volcanic eruption

7.1. INTRODUCTION

When we consider the fundamental resources needed for human survival, it is air and water that immediately come to mind. Soil is another fundamental resource that is essential for humans to survive on planet Earth. Sadly, it is yet another **natural resource** that's being exploited for economic gains with no thoughts of the future. Land pollution affects soil, which in turn affects agriculture. Agriculture and soil are the only things that are sustaining living beings on this planet. All the food chains in our environment have plant life as their source. Moreover, fertile soil is taken for granted and exploited like there is no tomorrow. It is increasingly used for agriculture with scant regard for and in many areas soil has turned almost barren. The reasons are manifold, including continuous farming of the same crops, rapid industrialization, and increasing development. Soil is home to various flora and fauna. It is a habitat for countless organisms.

Furthermore, the soil is also rich in nutrients. Any disturbance in this delicate system of flora, fauna, nutrients, and habitats adversely impacts the soil ecosystem. Moreover, any changes in the balance of soil ecosystem lead to problems with agriculture as is witnessed worldwide today. This is like a domino effect, which then further leads to environmental problems and the plunging economy. Hence, it's safe to say that soil and organisms living in it are an important natural resource and a part of our ecosystem and environment.

On the other hand, due to wind, rain, landslide, and water, soil erosion is a natural process which can't be prevented. Human intervention is causing irreversible damage to the soil. Mismanagement by humans is, therefore, causing soil erosion, which is totally avoidable. This leads to poor soil quality as the fertile top layer gets lost in erosion.

Other adverse effects are salinization, reduction in organic matter and nutrients in agricultural soil, and soil contamination. Such poor soil has very low agricultural productivity. In fact, it has been seen that in some cases agricultural growth has stooped to such a low extent that yield of seeds is less than what was planted. It is also not economically viable to support barren soil along with the living organisms that are dependent on it.

All human beings and animals depend upon agriculture, forests, and natural vegetation for sustenance, which in turn depend upon the soil. These are the most evident benefits of soil. There are other benefits too. Soil is an important part of the water cycle because it regulates run-off, i.e., the water which moves along the surface and is not absorbed by the soil. Soil is also an important part of the cycling of elements like carbon and nitrogen. Therefore, it acts as a natural sink for many contaminants like an urban waste, industrial waste, and natural pollutants, thereby protecting our environment from pollution. For these reasons, continuous and judicious management of soil is important. According to some estimates soil of almost 20 million square kilometers or 15% of the world's land area has been contaminated.

Natural resources are resources that exist without actions of humankind.

The main reasons for these alarming figures are deforestation, overgrazing, and poor agricultural practices. As per a recent report on Europe, the soil is being degraded there at an alarming rate, and it is being irreversibly damaged. The major problems that soil is facing are erosion, slope stability, local, and diffuse contamination, soil sealing, and soil acidification. Contamination of soil is going to continue world over if concrete steps are not taken to conserve it at the earliest.

7.2. THE SOIL RESOURCE

Soil is the thin layer covering our Earth. It formed over a very long period of time due to slow interaction of minerals, organic matter, living organisms, air, and water. The type of soil varies from place to place. The type of soil and rate at which soil is formed depend on factors such as geology, climate, vegetation, and relief of a place (Figure 7.1).

Figure 7.1: Selecting the right kind of soil for the work to be done is essential.

Source: https://cdn.pixabay.com/photo/2015/05/14/02/22/soil-766281_960_720.jpg

Biology is the natural science that studies life and living organisms, including their physical structure, chemical processes, molecular interactions, physiological mechanisms, development and evolution.

There are many characteristics of soil like texture, structure, **biology**, pH, salinity, etc. Many important ecosystem functions such as soil fertility and the transformation and degradation of pollutants depend upon these properties. During biogeochemical cycling of elements like carbon and nitrogen, soil organisms like fungi, bacteria, and earthworms play a very important role. As per some estimates, 25 tons of soil organisms can be found in the top 30 cm of one hectare of soil. Furthermore, it is being discussed how fossil fuels cause greenhouse effect and how they cause various types of air pollution Moreover, photosynthesis is used to provide fuels such as vegetable oils (e.g., biodiesel), wood, and other biomaterials (e.g., biopolymers). This has been compared with the use of fossil fuels in terms of the impact on atmospheric composition and sustainability.

Thus, once Earth's limited fossil fuel reserves run out, soil would act as a fallback option by providing biofuels and bio-feedstocks to industry.

The last part of this unit details the depletion of soil through erosion (due to wind and water), and through chemical and ecological degradation by various forms of soil misuse. Degrading and mismanaging soil leads to degradation of the environment. Therefore, it provides resource cycling in the maintenance of the biosphere. Soil acts as a buffer against changing environments (day-to-day and seasonal fluctuations in temperature).

Soil is teeming with diverse flora and fauna. There are different kinds of soils at different places too. Some have just the right mix of physical, chemical, and biological characteristics. On the other hand, others might not have just the right combination of these characteristics. This is the reason why we see a difference in productivity in gross terms in one soil sample from the other.

This clear-cut opinion of one soil sample being healthier than the other one and hence being more productive gets complicated if we consider the broader perspective. Consider this; soils with low levels of nutrients can have highly diverse flora and fauna with high levels of endemism. They might account for some pretty unique vegetation found nowhere else in the world. Thus, we can say that while accounting for soil resources, many dimensions need to be considered. Loss of soil through natural processes like erosion can be measured in terms of soil volume and in the nutrients and biological resources that are redistributed or destroyed. However, need to account for our soil resources in terms of their types, volume, nutrient content, and other characteristics for an extensive examination of agricultural and forest productivity and the impact of changing land uses.

Soil is an important natural resource and an indisputable part of the environment. Soil is the sole source of nutrients and water for vegetation, be it in forests or farms.

7.3. CONTAMINATION OF SOIL AND THE ENVIRONMENT

There is an extensive usage of chemicals in day-to-day lives. These chemicals contaminate the soil and water bodies in the process of gathering, transporting, processing, storing, using, or disposing. Among the contaminants are consumer products, fertilizers, pesticides, heavy metals from sewage sludge, solvents from cleaning fluids, and paints. Moreover, they tend to accumulate in the soil or leach into the groundwater. Some contamination activities are deliberate while others maybe involuntary, but they all lead to excess accumulation of pesticides, wastes, or other contaminants in the soil. Example of involuntary **contamination** can be accidental spill during oil production or the transport or storage of fuels or chemicals.

7.3.1. Acid Rain

In highly industrialized and populated areas, the air is saturated with Sulfur dioxide and nitrogen oxide. These harmful chemicals combine with water molecules in the air. Then they come down as rain which is called acid rain, aptly named as it's harmful for soil and all living things. This rain acidifies the soil, which causes essential nutrients like calcium and magnesium, to leach from the root zone and results in low agricultural output (Figure 7.2).

Contamination is the presence of a constituent, impurity, or some other undesirable element that soils, corrupts, infects, makes unfit, or makes inferior a material, physical body, natural environment, workplace, etc.

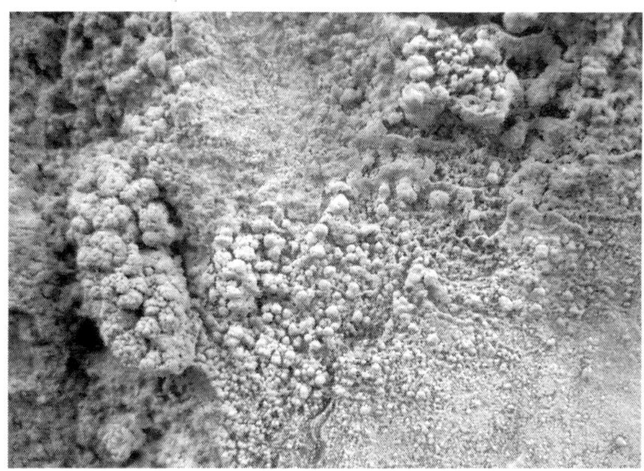

Figure 7.2: Acid rain is that one factor which leads to soil degradation.

Source: https://live.staticflickr.com/3227/2813673801_efe985d50f_z.jpg

7.3.2. Mine Lands

Mining for coal, metals, sand, gravel, etc. is important for the development of different countries. There are two types of mining-underground and strip mining. In general, we can say that indiscriminate mining of both kinds has left large areas covered with mine wastes and huge craters. Earlier, it was thought that if such areas would be left to themselves without being leveled, vegetation would ultimately reclaim them, but nature doesn't work this way. Reclamation efforts are time and resources consuming, and they can be non-economical in the long-term for the countries where the mining occurs. Recent studies in Guyana, a country with high gold mining industry in areas with important tropical forests show that it is more economical for the country to prefer REDD+ alternatives to be paid for non-deforestation and degradation of their forest than the cost of mining vs. benefits (Figure 7.3).

Figure 7.3: Alluvial gold mining in the rainforests of Guyana.

Source: https://blog.conservation.org/2017/01/guyana-aims-to-shift-economy-from-gold-to-green/

In some cases, companies doing reclamation indicate that pyrite present in the soils don't let any future vegetation to grow. Due to the same reason, such land can't be reclaimed for other activities too. Pyrite is known to release iron, sulfuric acid, and other metals when exposed to air. For United States regulation is now a federal crime to leave a mine site unleveled. Each mine site is unique, and reclamation of the site should be in a judicious manner to bring back the soil and vegetation. However, the damage of the mining to the terrain and elimination of soils, leaving only the rocks make impossible the rehabilitation and less possible the restoration of the original ecosystem.

7.3.3. Heavy Metals

There are metals like cadmium, lead, zinc found naturally in soil in very low amounts. These are needed for proper growth of plants and are not toxic to animals which consume these plants or organisms which live in soil. However, if the concentration of these elements increases beyond a limit, it becomes toxic for plants and animals.

How does the concentration of metals increase in the soil?

Reasons can be sewage sludge, mine tailings, pesticides, smelters, **metal** processors, leaded petrol, etc. These are all outcomes of rapid industrialization with little or no regard to the environment. In fact, in some areas, the metal concentration in soil is so bad that no vegetation can grow there. Those areas have become toxic to residents. Moreover, living in such areas is bound to result in health problems. EPA, the Environmental Protection Agency is working hard towards remedying this situation. It doesn't allow sewage sludge with excessively high metal contents to be spread on the soil. Sludge with lower concentrations is still allowed to be spread on the land with a strict check on the overall metal concentrations. Therefore, modern air quality standards have greatly reduced the amount of metallic pollutants in the air. This is a step forward in soil conservation.

Metals are opaque, lustrous elements that are good conductors of heat and electricity.

7.3.4. Waste Disposal

Over centuries soil has absorbed human waste, dead plants, animals, insects, organic matter and converted it to topsoil. This process is natural and has been utilized by humans over the years to dispose of waste. In fact, a hectare of land has the ability to degrade tons of human waste, dead plants, animals, insects, and organic matter (Figure 7.4).

Figure 7.4: Waste disposal in soil reduces the quality of soil, making it less fertile for various purposes.

Source: https://c.pxhere.com/photos/7d/15/photo-195734.jpg!d

After the industrial revolution, even industrial waste from different kinds of industries is being absorbed into the soil. However, if done properly with restraint and while following all environmental norms, these practices can be safe. These waste products should not contain excessive concentrations of salts or toxic metals. Disposal of industrial waste in such a manner can be done regularly without disturbing the soil ecosystem and without contaminating the groundwater.

7.3.5. Radionuclides

Radiation can be found everywhere in our environment. It has been present in Earth since the time it was formed when radionuclides originated from that process. These radionuclides have a long life, and so do the daughter radionuclides. Radiation also comes from cosmic ray interaction within the atmosphere. Human bodies also have some naturally occurring radionuclides, for example, potassium-40. Even home appliances like TVs, some smoke detectors and radioluminescent paints produce radiation. Moreover, Uranium, which is a metal occurring, naturally produces radon gas, which is again a source of radiation. This radon gas, according to WHO is a major source of radiation exposure at homes and in offices. In developed nations, **medical exposure** to radiations is also significant. There, medical exposure to radiation might even be more than the indoor exposure to radon gas.

Medical exposure differs from occupational and public exposure in that persons (primarily patients) are deliberately, directly and knowingly exposed to radiation for their benefit.

Nevertheless, there is the highest concern regarding exposure to radioactivity that comes from accidental release of anthropogenic radionuclides. With great advancements in nuclear technology come great risks. This is the age of nuclear reactors, nuclear weapons, and even nuclear medicines. Detonation of nuclear bombs in the past has resulted in radioactive fallout resulting in millions of deaths and contamination of soil. Generations of people have borne the brunt with genetic and other diseases (Figure 7.5).

Figure 7.5: Radionuclides.

Source: https://media.defense.gov/2017/May/11/2001745882/-1/-1/0/170413-F-DB515-0161.JPG

On the other hand, a nuclear meltdown is a nuclear accident which results in the release of nuclear material in the environment and contaminates the soil. Same was seen in Chernobyl, in 1986. Soil contamination after such events is not localized. This is due to the fact that radionuclides can travel great distances by becoming airborne. They can even travel thousands of miles. Furthermore, contaminated soil results in contaminated groundwater, plants, animals, and dust. Humans get exposed to radiation through them. This can result in an increase in genetic defects, cancer, birth defects, among other diseases in humans. Radionuclides are unstable and have excess nuclear energy, but they are not chemically different from their stable isotopes example tritium is a radionuclide of hydrogen.

There are many elements in nature that get absorbed by the human body. In the case of absorption of isotopes and radioisotopes, the same characteristics and processes apply that determine the fate and internal accumulation of other elements in the human body. However, there is a significant difference in exposure to chemicals as compared to exposure to radionuclides. Exposure to chemical contaminants occurs only at the site of contact (i.e., target organs). Exposure to radionuclides is both internal and external. Internal exposure results from the absorption of radionuclides in tissues and passage through the alimentary tract during digestion. It is similar to exposure caused due to chemicals. Thus, External exposure to radionuclides occurs by coming in direct contact through contaminated soil, sediments, or water. The amount of radiation absorbed depends on many factors like nature and number of radionuclides, the spatial association of the **organism** and the source, the characteristics of the media, and the location and size of the organisms. In case of exposure to many radionuclide sources over a period of time does is considered to add up and, in such cases, total exposure to radiation is the sum of internal and external doses for all radionuclides.

Organism is any individual entity that propagates the properties of life.

7.4. HEAVY METALS AND SOIL CONTAMINATION

The terms 'Heavy metals' and 'trace elements' are both used to describe elements with metallic properties. The word 'Heavy' in 'heavy metal' is misleading and not scientifically accurate. Definition of 'heavy' is not fixed. In fact, some heavy metals like arsenic and antimony are semi-metals or metalloids. Moreover, trace elements are those elements which are found in minute quantities in rocks and soil. They are present in varying amounts in soil depending on its location and the rocks that have broken down to make the soil's components.

The only elements that need to be considered under 'heavy metals' while taking human health risks into account are -Arsenic (As), Lead (Pb), Cadmium (Cd), Chromium (Cr) (only the form Cr (VI) is toxic), Copper (Cu), Mercury (Hg), Nickel (Ni) and Zinc (Zn).

Most of these elements are already consumed by humans in trace amounts in food or as supplements. They are commonly referred to as minerals like Iron, Magnesium, Phosphorus, etc. These are necessary for human health and cause deficiency diseases if not taken in appropriate quantities. At the same time, other elements like arsenic, lead, and mercury are poisonous for humans even in little amounts. Thus, the soil absorbs all these elements, poisonous or not, that are released in the air, water, and land due to industries or otherwise.

This feature has always protected the wider environment. As per Morgan, 2013 such a soil itself becomes a problem area as the crops grown on it are unfit to eat. Some areas naturally have high levels of heavy metals without even human intervention. Only that vegetation takes roots there which have the ability to thrive in such conditions. Industrialization can hardly be blamed alone for soil contamination with heavy metals. Even human activities contribute to this. Furthermore, some of them are mining, smelting, industry, agriculture, and burning fossil fuels. Other factors that contribute are dumping of leaded paint, electronic waste, and sewage containing heavy metals. However, there have not been studies focused on how **heavy metals** in soils lead to health problems in humans, although there have been many studies showing the effects of polluted air and water on human health. This report's focal point is four heavy metals identified in the WHO 'ten leading chemicals of concern' list: arsenic, cadmium, lead, and mercury and their impact on human health.

Heavy metals are generally defined as metals with relatively high densities, atomic weights, or atomic numbers.

Some pollutants are a result of human actions, and they impact land and water bodies. Such pollutants are known as biological pollutants. They include bacteria, viruses, molds, mildew, animal dander, cat saliva, house dust, mites, cockroaches, and pollen. They all have different sources which might include pollens from plants; viruses transmitted by people and animals; bacteria carried by people, animals, and soil and plant debris. Hence, heavy metals can result from both natural and human activities. They can ultimately end up anywhere either in soil, air, or water.

7.4.1. Natural Processes

Heavy metals are not just a byproduct of human activities. There are many natural sources of heavy metals too. They include volcanic eruptions, sea-salt sprays, forest fires, rock weathering, biogenic sources, and wind-borne soil particles. These natural activities which release heavy metals in atmosphere occur in different environmental conditions. Moreover, heavy metals can be found in soil, air, and water in the form of hydroxides, oxides, sulfides, sulfates, phosphates, silicates, and other organic compounds.

These heavy metals are found in traces, but they still have the ability to cause serious health problems to humans and other mammals.

7.5. SOIL EROSION AND ENVIRONMENT

The most common heavy metals are led (Pb), nickel (Ni), chromium (Cr), cadmium (Cd), arsenic (As), mercury (Hg), zinc (Zn) and copper (Cu).

Soil erosion can be defined as a process where the top fertile layer of soil gets transported to some other place. Foremost reasons for this are winds, water flow, rainfall, tsunamis, avalanches, and floods. Erosion is mostly a natural process, but it has been seen that human and animals' activities to contribute to soil erosion. Hence, when humans use land extensively for agricultural, it results in soil erosion. Consider the case where tillage is used to clear vegetation covering the soil while disturbing soil structures together with roots holding soil particles together. This causes extensive soil erosion. Deforestation and reduction of soil cover also lead to soil erosion.

On the other hand, the impacts of soil erosion are manifold, but the focal area of this report is how soil degradation by erosion impacts the food supply of the world. Soil is considered to be the world's most valued resource. "The loss of this resource, through land degradation processes such as wind and water erosion, is one of the most serious environmental problems we are faced with as it is destroying the means of producing food." Furthermore, food is a basic need without which no one can survive. Soil degradation impacts the world food supply. With erosion, top rich soil goes away, and it is this layer which is rich in organic matter, holds water, and affects the soil biota. "Rainfall's beneficial impacts are reduced too due to increases in water runoff and reductions in the soil's water holding capacity. Taken together or separately, these factors limit the soil's productivity and, as a result, can reduce crop yields from 15 to 30%."

As per a research carried out by the U.N around 11% of the world's best soils (about as big as India and China put together) have been spoiled by human activities since as early as 1935. "About 9 million hectares of arable land have been irreversibly damaged by erosion through overgrazing, deforestation, and unsustainable agricultural practices. A further 1.25 billion hectares is considered to be seriously degraded and could be restored only at great costs."

The world is full of examples and case studies where erosion has affected the world's food supply. Let's consider one by one:

It is clear that erosion leads to a drop in the world's food supply. Globally, 75 billion tons of soil is lost through deforestation as a major cause of soil erosion. As a result, there is a loss of about US$400 billion in a year (at US$3/ton of soil for nutrients and US$2/ton of soil for water), or approximately US$70 per person a year. It is estimated that the total annual cost of erosion due to agriculture in the US is about US$44 billion in a year or about US$247/ha of cropland and pasture. Further, in Sub-Saharan Africa the situation is worse. In some country's productivity has declined in over 40% of the cropland area in two decades and at the same time the population has doubled. Another major pain area of this region is overgrazing of vegetation by livestock as a result of which land becomes degraded.

7.5.1. Occurrence and Effects

Soil erosion happens when there is a lack of soil cover like rye, clover, etc., which have hardy roots that hold the soil in place. If the cover is not provided between crop cycles, then soil erosion keeps happening. This thus leads to all fertile soils being swept away, leaving behind barren soil unfit for cultivation or farming activities. However, due to the lack of vegetation cover, rainfall attraction does not take place, which may result in a famine. There are many countries stuck in exactly this situation. They depend on other countries for food supply. This global problem is brought about by ignoring the root cause, namely soil erosion.

On the other hand, the percentage of deserts in today's world is way higher than it was in the past. There is little or no crop production in deserts. Green areas can turn into deserts when there is a lack of vegetation. This is because vegetation attracts rainfall and acts as a soil cover for preventing soil erosion. Hence, when continuous soil erosion over a long period of time takes place, these areas lose their vegetation, which means there won't be water retaining capabilities or factors influencing rainfall attraction. What will be left is the soil exposed to wind and other factors that influence soil erosion. The main reason for this is deforestation. In due course of time, such an area becomes unfit for agriculture, which impacts food supply.

Research shows that as a result of soil erosion soil quality decreases. Food crops cannot be planted in such soil. In such places, those crops are planted which can thrive even in the poor-quality soil. Such crops are known **as cash crops**. They might provide money to the farmer but not food to the humans. They do not need many nutrients in the soil to grow. Furthermore, at the current rate at which soil erosion is happening, most of the world has become unable to produce food crops. Countries rely on other countries to produce food crops for them. This situation could not have come at a worse time because the world's population is at an all-time high, and previously food productive areas have become unproductive. However, soil erosion not just results in loss of fertile top layer, it also brings about reduction in crop growth, emergence, and yield. Water erosion can further remove seeds and plants from the eroded site. It is very easy for organic matter from the soil, residues, and manure to be carried off because they are lightweight like in spring thaw conditions. Thus, eroded soil may also carry pesticides along with it. As per Matende, 234 this has an adverse effect on the supply of food, more so when corrective measures are not taken in time.

Cash crop or profit crop is an agricultural crop which is grown to sell for profit.

"Wind erosion has also been seen to cause significant losses on food crops. The unavailability of lasting vegetation cover in a lot of areas has led to widespread wind erosion." "Loose, dry, bare soil is the most susceptible; however, crops that produce low levels of residue also may not provide enough resistance. Wind erosion may also create adverse operating conditions in the field." It can, therefore, lead to completely damaged food crops. This can lead to reseeding by farmers, which brings about delays. Another problem which wind erosion causes is due to the impact of saltation particles on the plant, often called sandblast damage. The end result being poor or no yield.

On the other hand, "Off-site impacts of soil erosion are not always as apparent as the on-site effects. Eroded soil, deposited downslope can inhibit or delay the emergence of seeds, bury small seedling, and necessitate replanting in the affected areas." It is water or rainfall that brings about such erosion. In heavy rains uncovered soil or one which is loose due to tillage and human activities can easily be swept away. This kind of erosion also results in little or no yield, but it has its other share of negative effects also. However, heavy rains lead to crops being buried, and some are left bare. Due to this, either germination will not take place, or the exposed seeds will be consumed by birds and other animals. The need of the hour is to take preventive actions else there will be little or no food production which will, in turn, affect the entire population of the globe.

REVIEW QUESTIONS:

1. Describe soil as a resource.
2. Why soil resources form a fundamental part of the environment?
3. How acid rains contaminate the soil?
4. Define Radionuclides.
5. Name some of the heavy metals that cause severe risks to human health.
6. What do you mean by biological pollutants?
7. Why soil erosion takes place?
8. Explain the consequences of soil erosion.
9. How waste disposal in soil can be an effective process?
10. Describe the impact of fertilizers and pesticides on the soil?

REFERENCES

1. *Accounting for Soil Resources,* (n.d.). [eBook] p. 3. Available at: https://unstats.un.org/unsd/envaccounting/seearev/GCComments/Chapter5-Australia2.pdf (Accessed on 18 June 2019).
2. Bharti, M. A., (2015). Soil health – an issue of concern for environment and agriculture. *Journal of Bioremediation & Biodegradation*, [online] *06*(03). Available at: https://www.omicsonline.org/open-access/soil-health--an-issue-of-concern-for-environment-and-agriculture-2155-6199-1000286.php?aid=50368 (Accessed on 18 June 2019).
3. Eugenio, N., McLaughlin, M., & Pennock, D., (2018). [eBook] *Food and Agriculture Organization of the United Nations*, p. 156. Available at: http://www.fao.org/3/I9183EN/i9183en.pdf (Accessed on 18 June 2019).
4. Loynachan, T., Brown, K., Cooper, T., & Milford, M., (2014). *Sustaining Our Soils and Society*. [eBook] American Geological Institute, p. 68. Available at: http://www.envirothon.org/files/2014/1_Sustaining%20Our%20Soils%20and%20Society%20(2).pdf (Accessed on 18 June 2019).
5. Masindi, V., & Muedi, K., (2018). Environmental contamination by heavy metals. *Heavy Metals*. [online] Available at: https://www.intechopen.com/books/heavy-metals/environmental-contamination-by-heavy-metals (Accessed on 18 June 2019).
6. Pimentel, D., (2006). *Soil Erosion: A Food and Environmental Threat*. [eBook] p. 19. Available at: http://saveoursoils.com/userfiles/downloads/1368007451-Soil%20Erosion-David%20Pimentel.pdf (Accessed on 18 June 2019).
7. Weil, R., (2002). *Soil and Environmental Quality: A Course for Nonmajors*. [eBook] p. 6. Available at: https://enst.umd.edu/sites/enst.umd.edu/files/_docs/weilsoilandenviroqual.pdf (Accessed on 18 June 2019).

Chapter 8

SAVING THE FUTURE, SAVING THE WORLD

LEARNING OBJECTIVES:

In this chapter, you will learn about:

- The requirement of environment security.
- Millennium development goals.
- Characteristics of a prevailing guardian.
- Dealing with the environment related issues in an effective manner.

Key Terms:

- Cooperation
- Global change mission
- Global ethic
- International relations
- Millennium ecosystem assessment
- Mitigation
- Neo-Malthusian approach
- The millennium development goals
- Védegylet

8.1. INTRODUCTION

Sustainable development or the development that caters for the current requirements without putting at stake the resources for the future generations has human health and environmental protection as its core values. Moreover, the interaction and dependence of health and global environmental change on each other has brought forth the importance to enhance the lives of the most vulnerable and poor sections whilst at the same time, preserving the **fragile ecosystem**. However, drastic changes to the ecosystem and climatic changes which have been the result of unplanned development in terms of its effect on the environment have finally shaken up the world and called for steps to be taken for environment preservation. An increasing effect of the environmental changes on human health has thus called for combined efforts across the global to tackle challenges like biodiversity loss, land degradation, and climate change.

Environment greater effects and is in turn, affected by the health of the nation. Owing to this, a greater concern on the environmental parameters may only help to garner support from the populace that can lead to ambitious and new global environmental policies. If this is not done timely, a large number of populations are likely to suffer new and greater health-related risks.

According to the Millennium Ecosystem Assessment, the wellbeing and protection of our future generations lie in careful planning and utilization of natural resources. This is largely dependent on how decisions are made. In the current scenario, such environment-related concerns can be relegated to a position of planning that fall after wealth creation or national security. Our current decisions are likely to have an impact on the quality of life of the children across the globe in 2050 (Figure 8.1).

The Eight Millennium Development Goals (MDGs) 2015, however, have not yet been achieved:

- to eradicate extreme poverty and hunger;
- to achieve universal primary education;
- to promote gender equality and empower women;
- to reduce child mortality;

- to improve maternal health;
- to combat HIV/AIDS, malaria, and other diseases;
- to ensure environmental sustainability; and
- to develop a global partnership for development.

Fragile ecosystems are important ecosystems, with unique features and resources.

Figure 8.1: Sustainable development would help in dealing with all types of environmental issues.

Source: https://www.maxpixel.net/static/photo/1x/Wind-Turbine-High-Tech-Sustainable-Development-4136758.jpg

In order to help people to escape poverty and lead a happy life, there has been a plethora of discussions and debates on the requirement for change in human development. The 1970s saw a lot of discussion on the sustainability aspect, and in 1992, international equity was recognized as central to policymaking that would help safeguard the future. This principle was accepted by the UN Conference on Environment and Development and has now, over a period of time found its way into the constitution of most countries. At the same time, there are no universally binding policies for its implementation. Thus, the short-term representative democracies of nations may be the main cause behind this aspect. The World Commission on Environment and Development (WCED) has categorically stated that "We borrow environmental capital from future generations with no intention or prospect of repaying. Hence, we act as we do because we can get away with it: future generations do not vote; they have no political or financial power; they cannot challenge our decisions."

Commissioner is, in principle, a member of a commission or an individual who has been given a commission (official charge or authority to do something).

Countries like Israel and Hungary have taken positive steps to deal with this institutional shortcoming. Both of them created independent voices in the form of a **commissioner** or guardian for the generations to come. These are more like temporal checks and balances in the bigger issues dealing with sustainability. Moreover, the commissioners have unlimited access to the information guiding policymaking with due respect to a sustained development (Israel) and the basic right of humans to a healthy environment (Hungary). Citizens' concerns are addressed by them, and at the same time, they reveal the long-term effects of current decisions to the public.

On the other hand, destruction of the environment, climate changes, financial crises, and the ever-increasing economic gap between the various strata of the society the world over is causing fear and insecurity amongst the masses. It thus needs to be acknowledged and accepted that these challenges are not stand alone or isolated. No country can handle them alone, and at the same time, due to their structural nature, there is an inherent requirement of up gradation in the available expertise in the field. Furthermore, the current generation has seen a drastic amount of changes and innovations in terms of communication, transportation, and overall development. It is hence ironic that the decisions made by this very generation alone shall affect the lives of millions to come. Sadly, the current thinking is stagnant and obsolete when it comes to the challenges of a growing population in a globalized world. With a persistent fear of spending too many resources and not being far-sighted enough, the current generation stays mired in an environment of non-action.

8.2. WHY IS THE UNFINISHED AGENDA: THE MILLENNIUM DEVELOPMENT GOALS (MDGS)

It is found that far to complete those goals inequalities seem to be persisting and rather rising in spite of a growth in the **global economy** since 1992. The poorest nations and their populations seem to be facing the biggest challenges in terms of health due to environmental challenges even though globally all nations are likely to face similar repercussions in the future. Moreover, studies conducted by WHO revealed that the health of populations in the poorer nations is already reeling with the negative effects of environmental changes.

Global economy is the economy of the humans of the world, considered as the international exchange of goods and services that is expressed in monetary units of account.

However, for populations already suffering due to malnutrition, poverty, and natural/human-induced disasters, the loss of ecosystem is only adding to the disparity in terms of the availability of relevant and adequate health services. This calls for immediate action globally to enable the developing countries to provide better health services and reduce its vulnerability towards environmental changes that are likely only to intensify. These requirements have been spelled out in The MDGs. However, these goals deal with the basic human rights that need to be provided to every human being irrespective of nation or region of its inhabitation. Hence, these deal with a world where environmental sustainability is given priority; where there is equality between men and women; there is freedom from extreme poverty and hunger; health and shelter are available to all; women have a right and say over issues related to childbirth and safe childbirth is possible; education is equally available, and a productive, safe employment is available. Even though the applicability of MDGs was till 2015, they continue to be the benchmarks for basic standards of human development and have stayed relevant till date.

However, MDGs can be achieved by making the social determinants of health working in tandem with the ecosystem approaches. For this, one needs to recognize and consider the factors that result in poverty and vulnerability, grossly wrong trade practices across the globe, and a lack of education. The upstream ecosystem determinants are then casually connected.

8.2.1. A Need to Guard the Future

The WCED was established by the UN General Assembly in order to carry out an in-depth study on the seeming discrepancies between developmental and environmental concerns. *Our Common Future*, the final report which came out in 1987, brought out the large amounts of ignorance and inertia that exists amongst the governments of various nations.

In a way, the report shook up the nations to wake up and take a call on sustainable development which seems to be the only way in which development can take place taking due care of environmental issues. Thus, the in-depth analysis made by the report clearly laid down that: "We have tried to show how human survival and well-being could depend on success in elevating sustainable development to a global ethic." It defined sustainable development as that which "meets the needs of the present without compromising the ability of future generations to meet their own needs." Therefore, inter-generational equity was considered the key principle in the UN Conference on the Environment and Development, the ensuing world conference in 1992. There was a consensus amongst at least 190 states that even as the nations utilize natural resources to meet their developmental needs, it cannot be at the cost of future generations being deprived of the use of the very same resources.

However, whilst pursuing the development and **enhancement** of our living standards, the need to bear in mind the long-term impact of such actions cannot be stalled. It is hence essential that development works hand-in-hand with sustaining the environment as well as the natural resources for the generations to come. Agenda 21 or an action plan to deal with sustainable development and the means to bring it about was agreed upon by the states.

Enhancement is any product change or upgrade that increases software or hardware capabilities beyond original client specifications.

However, some of these scholarly concerns and discussions found their way into initiatives like the Future Generations Program of the Foundation for International Studies at the University of Malta supported by UNESCO and the Institute for the Integrated Study of Future Generations in Kyoto, Japan founded by Katsuhiko Yazaki of Japan and Tae-Chang Kim of Korea. Moreover, a large number of essays and conferences during the 1990s were a direct result of these two. The Foundation for the Rights of Future Generations in Germany actively engages in the political scene as well by holding conferences and bringing out publications in both English and German. It not only raises awareness on relevant issues but also brings about publications in English and German. *Intergenerational Justice Review* is one of the many publications brought out by the foundation.

The effect on future generations due to development and the consequent environmental concerns had been a subject of ethical discussions even prior to 1992.

However, in spite of the global sustainability summit even after twenty years thereof, the implementation of the policies has not been effective. There have been scientific observations and measures that reveal the importance to act immediately on sustainability-related issues but still targets with respect to the protection of the oceans, climate change **mitigation**, biodiversity, social equity, and health and poverty eradication are far from being achieved.

Mitigation is the effort to reduce loss of life and property by lessening the impact of disasters.

In such a scenario, the current measures and the unwillingness to change the current means to deal with such issues becomes questionable.

These institutions are mired in an environment of the closed decision-making process where narrow and limited concerns are dealt with. Further, their independent existence and fragmented nature isolate them from those who are responsible for the utilization of natural resources and those responsible for ensuring a sustained environment. At the same time, institutions managing the economy are separate. All these independent and segregated institutions shall bring down the positive influence of integrated and interlocked ecological and economic systems. However, every individual whose rights and wellbeing is likely to be affected by a decision that has long-term ramifications should be given a voice in that decision making. This would further help in incorporating a system of checks and balances within a democratic set up that is structurally oriented towards a short term. Therefore, a system like having an ombudsman whereby the long-term concerns due to political decisions can be dealt with the need to be incorporated in the system. Direct involvement of the civil society would enable the ombudsman to have access to all governmental information thus increasing the possibility of having effective policies that shall be a direct result of the exchange of ideas at the early stages of policymaking. Thus, there is no congruency, and there is a balance between economic goals and resource regulation.

Our Common Future brings out certain relevant points. In the first place, the existing institutions are not equipped to deal with the challenges and issues that are relevant in the current changing times.

On the other hand, the guardians would be able to directly involve themselves in the decision making of various departments by explaining to them the effects of specific decisions, their effect on future generations in terms of the living conditions and the overall viability, cost-effectiveness of those decisions. In order to enable sustainable growth, the office of such an entity would become a storehouse of information and expertise that would work beyond just the immediate needs of GDP and include political goals, targets, and indicators that are all-encompassing.

8.2.2. Characteristics of a Powerful Guardian

For a guardian of future generations to be able to perform the function of keeping in place checks and balances, certain characteristics are essential. As per the basic concept of division of power, such a guardian should be independent. Being a legally independent office, the guardian should not hold any other governmental post (for instance being a part of the parliamentary committee). For instance, even though the Hungarian commissioner depends on the decisions of the Parliament for his budget, he enjoys immense independence. The guardian's decisions should then be legally binding so that they can maintain their efficacy. Currently, such an independent and legally binding position is reposed only in the Hungarian commissioner. However, deputies in a commission work based on a stung trust factor that brings about co-operation within the commission.

This very premise can be shaken in instances like Israel, where the commission enjoyed de facto veto power by way of tactical delivery of statements. Size of the office too affects the efficacy of the commission. There have been varied changes in different regions, for instance, the Hungarian office employs 35 people; as per the law, the Israeli commission staffs six people and over a period of time New Zealand has increased its staff strength from ten to twenty. Moreover, trust forms the very genesis of the guardian's office, and this requires it to have a clear mandate with transparent, regular reporting channels. However, of the three guardians mentioned above, even though all of them cater for regular reports, the Hungarian commissioner ensures their independent distribution through his direct mandate whereas when it comes to media relations, the opinion of both the commissioners of Israeli and New Zealand is influenced by either the legislative or executive bodies.

In order to function effectively, the guardian's office should have a strong public support, and it should be legitimate. Top-down approaches were used for the establishment of the offices in New Zealand and Israel. In Israel, media was used to communicate the work of the guardian and in New Zealand for all the cases investigated; good relations were directly built and maintained with all the stakeholders by the Commissioner's office. On the other hand, grass root initiative by a civil society organization called Védegylet (Protect the Future) led to the establishment of the Commissioner's office.

Access to information is one of the most important factors that enable the guardian's office to accomplish its tasks effectively. Unless the office has authority to request for relevant files and access them when required, its hands would be tied. In this aspect, the Hungarian commissioner has vast amounts of power and authority.

At the same time, it is equally important that the office allows for an inclusive and institutionalized input so that it is accessible for integrative assessments. Unlike Israel, both New Zealand and Hungary allow its citizens through the medium of petitions to have direct access to the ombudsman. As a result, the grassroots levels are strongly engaged in the work of the Guardian's office.

Depending on the cultural, political, legal, and regional requirements, each place is bound to have a different setup and rules for a guardian for future generations. Whereas the Israeli commissioner dealt with the living conditions of the future generations in a holistic manner under with twelve policy areas (which was closer to UNESCO's declaration dealing with future generations), the powers of the Hungarian and New Zealand commissioners were limited to matters dealing with environmental protection.

In lines with the aims of the European Union as defined in the Lisbon Treaty, a civil-society coalition has begun to take shape which promotes a guardian having a legal status similar to that of a constitution. Three aims have been listed in Article 3: "to promote peace, its values and the well-being of its peoples."

In order to achieve these aims, a wide range of objects varying from economics, culture, and security need to be explored. Within these parameters, a framework is built that helps review the policy decisions having a long-term effect on the wellbeing of future generations.

An EU level guardian is bound to bring about intergenerational solidarity as it would have the mandate to take a stand on behalf of future generations, it would support EU's measures that support sustainable development and enhance the coherence, efficiency of European policies drafted in single–issue departments. At the same time, a perusal of the limitations faced by a 20-person institution like the one in New Zealand is bound to ensure that the expectations from such a guardian are realistic.

8.3. HANDLING THE ENVIRONMENTAL PROBLEMS IN AN EFFECTIVE WAY

A range of issues and a large number of agencies involved in various environment-related issues are bound to make environmental problems tedious and complicated. Many times, the exact source of the problem is not identifiable making it further complex. A range of different but interrelated issues often affects a single environmental problem.

As such, no such issue occurs/exists in isolation. The human factor is at the core of most environment-related issues. For instance, pollution in any form is the direct result of human action. One was to curtail this would be to restrict or cease activities that are a direct cause of the harm that is caused to the environment.

Action plan is a document that lists what steps must be taken in order to achieve a specific goal.

Clear identification of the problem is at the root of finding a clear and plausible solution. Once the causes are known, it shall enable environmentalist and environment managers to deal with the problem head-on.

Once the problem is clearly an **action plan** is required to be put in place so that the relevant agencies and interested individuals can take up clear cut roles to help solve the issue. There should be an elaborate plan whereby each party remains focused and has clear directions about the steps to be taken. An initial testing of the water and soil, plant surveys, and wildlife inventories shall provide a baseline where the success/ failure of the plan can be gauged. This must be conducted at the initial stages of the action plan.

It is always wiser to find the exact cause of the problem. This enables placing timely and appropriate measures to deal with it. At times, the source of the issue may be clear, and at other times, it may not be as obvious. For instance, a local water source may be contaminated due to the presence of an abandoned mine close by that has acidic discharge. On the other hand, a runoff creating nonpoint source pollution (NSP) may not have any clear source.

To be able to find a plausible solution, it is essential to first find the causes by a simple process of elimination. In most cases, environmental changes are brought about by human action and a simple step like eliminating/ restricting human presence in such areas allows nature to recover. Many times, the source of the environmental issue may be far from the actual place where its impact is felt.

Re-testing and re-survey provide the information with respect to the efficacy of a measure taken. Sometimes the recovery may be slow and not evident at the initial stages, but the levels thereof can be a measure of the reduction in the adverse effects.

The existing environment laws need to be explored to check for any violations thereon. For instance, a state or federal law may be handy in dealing with violators where the industry is the cause. Many a times, the legislation may not have an inbuilt mechanism to deal with the relevant issues as in the Clean Water Act of 1972 wherein there is no provision to deal with agricultural runoff. In such cases, effort needs to be made to contact legislators so that relevant laws and regulations can be created. Another limiting factor is cost, which tends to go high for cleanup activities.

This may require additional resources and funds. Individuals, the general public needs to be educated about the harmful effects on the environment that their actions are likely to cause. Most of the environmental issues are a direct result of non-recognition and acceptance of the effects of human action on the environment. Over a period of time persistent littering and failure to reuse, recycle has only augmented environmental concerns.

8.4. THE PATH FOR COOPERATION: WHY GLOBAL ENVIRONMENTAL ISSUES "BELONG" TO INTERNATIONAL RELATIONS

As the world is growing, it is becoming more and more compact. Distances have reduced, economies have expanded and become more and more interdependent. Environment and related issues are one such aspect which is of equal concern to all nations across the global. It is perhaps one of the few factors that have ramifications beyond the boundaries of a single nation.

The science, technology, and multiple disciplines involved in environmental issues in different countries are all inter-related and interdependent. It is equally important for nations to develop strong scientific bonds with respect to environmental matters.

An interdisciplinary approach and understanding of the various parameters affecting future evolution is an essential pre-requisite for a sustainable growth, especially when it comes to an environment which is essentially a global issue. In order to understand the global implications of regional and local actions common paths need to be framed that can become a guiding factor for stakeholders all over. At the same time, the risks involved should be deduced so that international co-operation can be on fair and transparent grounds.

In the current times, the world is shrinking and coming closer, actions in one corner of the world have ramifications in other parts of the world. In such a scenario, international relations are the most interdisciplinary and relevant discipline. To avoid chaos and disorder with respect to precious natural resources and our fragile environment, it is essential that action is taken globally through co-operation and positive growth.

Environmental degradation is the deterioration of the environment through depletion of resources such as air, water and soil; the destruction of ecosystems; habitat destruction; the extinction of wildlife; and pollution.

The wide-reaching scope of International Relations in terms of developing strategic relations between states, cross border transactions and its multi-faceted approach towards various global issues, it becomes the aptest means for developing collaborative and effective solutions for preventing **environmental degradation**. Being a repository for political, social, and economic dynamics the world over it can help achieve collective response and action at the international level.

Globalization has resulted in a situation where actions at a regional level have global ramifications and vice versa. In such a changed scenario, collaborative effort, and an agreement on the basic idea of a concerted effort towards myriad problems including environmental issues seems to be the only way to move forward. Adapting to this new way of thinking and thereby, acting on such premise shall result in a peaceful and effective solution to the various challenges faced by the current times.

Owing to its creation post the First World War, International Relations is a relatively new concept, and a large number of countries, especially in the developing world, do not have the requisite scientific knowledge in terms of skilled manpower, resources, and methodologies that can help them deal with the current challenges.

Within itself, this becomes a major obstacle for the growth of International relations in such countries. A major challenge to the scholars in this field is to prove and convince its worth to such nations and thus develop the discipline in the changing world with interconnected and interdependent issues. Steps towards developing the discipline shall have to be interlinked with scientific progress amongst scholars in this field so that even as individuals understand the world as it exists today, there can be common paths towards decision-making and their eventual result.

As a result, it would be possible to both evolve new policies and keep the general public better informed so that they can contribute towards building up a specific thought or idea in a country thus moving towards the concept of global citizenship.

Water is one of the most abused of the various naturally occurring resources. If water management is tied up with global governance, there can be remarkable improvement with respect to handling issues like population growth, economic development, and climate change that are causing pressure on this precious resource.

Human rights are rights inherent to all human beings, regardless of race, sex, nationality, ethnicity, language, religion, or any other status.

In 2014, the Calouste Gulbenkian Foundation found that water is the second most important resource for producing food and across the world about 30–50% of the food produced goes waste. There is immense pressure on this vital resource which has a direct impact on **human rights** and the very existence and growth of vulnerable regions around the world.

The world is currently facing a large number of challenges, which are a direct result of civilization and growth. To bring about a sustainable growth, the international community needs "re-engineer the energy of the nations." Leaders and citizens of the world need to come up with practical and committed solutions where the world is supported by a green economy with optimum utilization of the available resources.

Awareness of the issues alone cannot suffice when it comes to actual commitment and concrete steps that require global co-operation and living up to common responsibilities and goals. The action taken by nations needs to go beyond the narrow walls of national boundaries. The existing neo-Malthusian approach towards the future may not suffice when it comes to facing the challenges of the ever-changing world with its myriad facets. For a concrete resolution to the environmental woes, the changing times and new opportunities that they offer need to be considered in a holistic manner.

8.4.1. Recommendations for the National Research Program

An understanding of global environmental change in all countries is needed, and nowadays all countries in the globe are working to reach the Millennium Goals by 2020, and many of them are also involved interdisciplinary collaboration projects between the natural, social, and behavioral sciences. Most of the countries today are fully advocated to generate governance and profound developmental changes towards the major environmental issues facing humanity as it is climate change, but also pollution by plastics. Most of the environmental issues are interconnected as it is deforestation with climate change, water resources, and food. Finally leading to change in populations, migration, wars, extreme poverty, and health in human populations.

REVIEW QUESTIONS:

1. What is the need for environmental protection?
2. Define Millennium Development Goals.
3. What are the characteristics of a powerful guardian?
4. How can environmental issues be tackled in an effective manner?
5. Describe steps involve in handing environment-related issues.
6. Why global environmental issues “belong” to International Relations?
7. Explain recommendations for a national research program.
8. Which factors are spreading fear and insecurity all over the world?
9. What do you mean by sustainable development?
10. Why is it important to educate the public regarding the environmental impact of hazardous substances?

REFERENCES

1. Everything connects, (n.d.). *The Environmental Crisis*. [online] Available at: http://www.everythingconnects.org/the-environmental-crisis.html (Accessed on 18 June 2019).
2. Göpel, M., (2010). Guarding our future: How to protect future generations – the solutions journal. [online] *The Solutions Journal*. Available at: https://www.thesolutionsjournal.com/article/guarding-our-future-how-to-protect-future-generations/ (Accessed on 18 June 2019).
3. https://www.who.int/topics/millennium_development_goals/about/en/
4. Pereira, J., (2015). *Environmental Issues and International Relations, a New Global (dis)Order – the Role of International Relations in Promoting a Concerted International System*. [online] Scielo. Available at: http://dx.doi.org/10.1590/0034-7329201500110 (Accessed on 18 June 2019).
5. Rogers, C., (2017). *How to Handle Environmental Problems*. [online] Bizfluent.com. Available at: https://bizfluent.com/how-5711846-handle-environmental-problems.html (Accessed on 18 June 2019).
6. Watts, N., Maiero, M., Olson, S., Hales, J., Miller, C., Campbell, K., Romanelli, C., & Cooper, D., (2019). *Our Planet, Our Health, Our Future* (p. 64). [eBook] World Health Organization. Available at: https://www.who.int/globalchange/publications/reports/health_rioconventions.pdf?ua=1 (Accessed on 18 June 2019).

INDEX

N

O

P

R

S

T

U

W